Aero Detail 24
NAKAJIMA Ki-84 "FRANK" HAYATE

カバー・表紙イラスト／野原 茂
Cover Illustrations / Shigeru Nohara

AF731772

Photo provided by Shigeru Nohara

CONTENTS

唯一の現存『疾風』製造番号1446の軌跡

History of Ki-84 Ko S/N 1446

解説／野原 茂
Commentary by Shigeru Nohara

敗戦という現実の前に、戦争を生き抜いた旧日本陸海軍機の多くも、連合軍の厳しい処分命令により、その大半が焼却されたり、スクラップとなって消えた。今日、世界各地の博物館に現存する数少ない旧日本陸海軍機は、その意味できわめて幸運な機体といえる。

この現存日本機のなかで、ドラマチックな奇跡を辿ったという点で、今回のエアロ・ディテールNo.24の主役となる四式戦『疾風』、製造番号1446の右に出る機体はないであろう。そのすべてを記述できるほどの資料は、残念ながら持ち合わせてはいないが、断片的なものをつなぎ合わせれば、その概要は紹介できる。本書の主旨からすると、ヒストリー的なページはさほど必要ないかもしれないが、この貴重な現存機の歩みを知っておくだけでも、その価値を理解する度合がちがうだろう。

キ84甲／製造番号1446(以下、単に1446号と記す)は、昭和19年(1944)夏から秋ごろにかけて、群馬県太田市に所在した、中島飛行機(株)太田工場にてロールアウトした。そして、飛行第11戦隊第2中隊に配属され、フィリピン決戦参加のため、ルソン島に進出したのであるが、11戦隊は、これに先立つ19年10月中旬の台湾防空戦で、隊長以下大半の戦力を失う損害を出しており、最初の進出時はわずか数機にすぎなかった。

その後、生存操縦者約10名が内地に帰還して、茨城県の下館飛行場にて戦力回復を行ったのち、19年12月17日、ふたたびフィリピンに進出し、末期段階に入った航空戦に参加した。こうした経緯からして、1446号は、この2回目の進出時の1機だった可能性が高い。

しかし、圧倒的に強大な米軍戦力を相手に、11戦隊もほかの四式戦部隊と同様、半月の間に機材の大部分を失ない、20年1月には残存機も特攻機として抽出され、わずかに生存した操縦者、地上員が、同月下旬に台湾へ、ついで内地に帰還した。

1446号は、この間エンジン不調、もしくは機体の損傷によるものか、ルソン島のクラーク飛行場にとり残され、20年3月、進攻してきた米軍により、同地でほかの多くの陸海軍機とともに接収された。

四式戦は、当時の米軍にとっては敵の新鋭戦

During the latter stages of WWII, well-trained Allied pilots decimated the ranks of Japanese aircraft. In the final, desperate stages of the conflict, the Japanese themselves launched their planes on one-way suicide missions. And what few aircraft managed to survive this carnage were treated no less severely by the American occupiers of Japan after the war. Melted down or bulldozed into giant pits, the remains of Japanese air power were quickly eliminated. As a result, painfully few examples of Japanese aircraft from WWII remain today. While a handful of their most vaunted fighter, the A6M Zero, still exists, nearly all other major types are either extinct, or limited to a single survivor. That's what makes the world's only surviving Ki-84 Hayate such a valuable aircraft.

The protagonist of this volume of the Aero Detail series is the Army Type 4 Fighter (Ki-84) Hayate, serial no. 1446 (referred to as just "1446" below). In addition to being the only remaining Ki-84, 1446 may have the most dramatic background of any surviving Japanese WWII aircraft in the world. While the focus of our series is not the history of aircraft, but their structure, knowledge of the background of this particular airframe can only help one to appreciate its historical value.

1446 was rolled-out in the late summer or autumn of 1944 at Nakajima Hikoki's Ohta factory in Ohta, a small city in southern Gunma prefecture, about 60 miles north of Tokyo. The plane was shipped to Luzon in the Philippines where it was to be used by the 11th Sentai, 2nd Chutai, but that unit suffered severe losses in Formosa in October and their return to Luzon was delayed. They finally arrived back in the Philippines on December 17, 1944 and joined in the final air battle in defense of the islands. At this point, it is very likely that 1446 was used in combat by the 2nd Chutai.

The Japanese air forces were ravaged by their superior American foes and their air power in the Philippines was nearly eliminated in the span of about two weeks. Beginning in January of 1945, surviving aircraft began to be used in kamikaze attacks, and later in the month, the few remaining pilots and ground crew returned to Japan.

◀1945年3月、フィリピンのルソン島クラーク飛行場に放棄されたまま、米軍に接収されたときの、1446号機。尾翼に描かれた白フチどり赤の雷光マークが、飛行第11戦隊第2中隊所属だったことを示している。方向舵下部に記入された"46"(白)は、製造番号下2桁。

1446 abandoned at Clark Field in the Philippines in March of 1945. This was the condition in which U.S. forces found the aircraft. The red lightning bolt with white border marks the plane as being of the 11th Sentai, 2nd Chutai. The "46" marked on the fin is the last two digits of the serial number.

▶オリジナルの塗装をすべて落とされ、入念な整備、補修ののち、T.A.I.U.(Technical Air Intelligence Unit)—S.W.P.A.の手でテスト飛行される1446号機。キ84のプロフィールを鮮明に示す、資料製の高い写真である。このあと、本機は米海軍小型空母ロングアイランド(CVE-1)に積み込まれ、米本国に運ばれた。

All original paint having been stripped away, 1446 is seen undergoing flight tests as the hands of the T.A.I.U.—S.W.P.A. This exceptionally clear photo shows of the lines of the Hayate extremely well, even if the Stars and Bars on the fuselage do make for an odd atmosphere. Shortly after these tests, the plane was loaded aboard CVE-1 *Long Island* for its trip to the U.S.

前頁：富士重工業(株)宇都宮南工場に移管されていた、1976年2月当時のスナップ。入間航空ショーにおける里帰り飛行から2年4ヵ月が経過していたが、この時点ではまだ充分に飛行可能なコンディションを維持していた。

Facing page : 1446 seen outside Fuji Heavy Industries' Utsunomiya South plant in February of 1976. Nearly 2-1/2 years had passed since its flight at the Iruma air show, but the plane still looked flyable at this point.

キ84甲 諸元 Ki-84 Ko Data

全幅(m) Wingspan(m)	全長(m) Length(m)	全高(m) Height(m)	主翼面積(㎡) Wing area(㎡)	自重(kg) Weight(kg)	常装備重量(kg) Loaded weight(kg)	エンジン名称 Engine	離昇出力(hp) Take-off hp	プロペラ名称 Propeller	最高速度 Max.speed (km/h/m)	上昇力 Climb speed (m/分秒)	実用上昇限度(m) Service ceiling (m)	航続距離距離(Km)+戦闘時間(分) Range(km + min)	燃料容量(ℓ)(標準) Fuel capacity(ℓ), stan.	燃料容量(ℓ)(最大) Fuel capacity(ℓ), max.	武装 Armament	搭載兵装 External stores
11.238	9.740	3.385	21.00	3,660	3,613	ハ45-21 Ha45-21	2,000	ペ32 Pe-32	624/6,500	5,000/6'26"	10,500	1,000 + 30	697	1,097	12.7㎜ MG x 2 20㎜ cannon x 2	250kg bomb x 2

闘機だけに、最重要調査、テスト対象機となり、1446号機は、さっそく専任のT.A.I.U.－S.W. PACIFIC—航空技術情報隊・南西太平洋地域—の手によって、飛行可能な状態に補修・整備された。

さいわい、本機の損傷、もしくは故障程度は軽く、他の捕獲機の部品を流用するなどして短期間に飛行可能となり、"S17"の登録記号を付与され、最小限の飛行性能テストが実施された。上の写真は、その際に記録用として撮影されたうちの1枚である。その後、本機は米海軍の小型空母に積載され、オーケストラ経由で米本国に運ばれた。

なお、この1446号機のほかにも数機の四式戦が戦後米国に送られており、陸軍航空隊の主要基地ライト・フィールドなどで、さらに綿密な調査、テストをうけている。1946年4月から5月にかけて行われた、このテスト項目の中には、米陸軍航空隊のレシプロ戦闘機、P-47N、P-51Hとの性能比較も含まれており、テスト機は最高速度こそ両機に比べて見劣りしたものの、初期上昇率、運動性に勝り、操縦性も良好、航続距離はP-51Hとほぼ同じという結果をのこした。

戦時中の日本陸軍では、せいぜい87オクタンのガソリンしか使えず、ハ45エンジン自体の不調もあって、最高速度はカタログデータの624km/hはとても出ず、580～590km/h程度が精一杯だった。それが、米軍の140オクタン価というハイ・グレード・ガソリンを使用したせいもあり、ハ45エンジンは調子もよく、カタログデータ以上のパワーを出したため、テストでは689km/hという、日本側のデータを60km/h以上も上まわる高速を出したのである。機銃、機関砲の弾丸なども積まない、軽荷重状態ということもあった。こうした結果をうけ、米陸軍の

At this point, 1446 was left behind at Clark Field, having been spared because of engine trouble, combat damage or some other reason. In March, the plane was captured along with many other IJA planes by advancing American forces.

At the time, the Americans correctly regarded the Hayate as their enemy's most capable and advanced fighter aircraft, and it was given high priority for research and testing. As such, for this purpose 1446 was quickly singled out for restoration to flyable condition by the T.A.I.U. – S.W. PACIFIC (Technical Air Intelligence Unit, Southwest Pacific).

Fortunately the plane's damage or trouble was light, and with the use of parts from other orphaned Ki-84s, the plane was quickly able to once again take to the skies. At this time it was given a registration number of "S17" and subjected to a few, short flight tests. The photo above was taken during one of those flights.

Loaded onto the escort carrier USS *Long Island*, the plane was then shipped back to the U.S. where it was placed into storage for the proposed National Air Museum. The following spring, another captured Hayate (believed ultimately scrapped) went through more rigorous testing at the Middletown Air Depot in Pennsylvania. Taking place in April and May of 1946, to their great surprise, the Americans found that in many performance categories, the Ki-84 was the equal of, or superior to, the P-47N and P-51K to which it was compared. Roughly the same in maximum speed at altitude, the Hayate proved superior in early climbing performance, maneuverability and overall handling. On the down side, the testers noted the aircraft's relatively short range, and poor protection, as well

◀米国で性能テストをうけたキ84甲。垂直安定板に記入された"302"は、テスト機の登録コードである"T2-302"を示す。識別写真撮影のため、上面ダークグリーン、下面グレイの、旧日本陸軍当時の迷彩に似せた塗装がほどこされている。本機は宇都宮で接収され、1945年11月3日に横須賀から米国に運ばれた2機のうちの1機で、P-47N、P-51Hとの比較審査に用いられた。

TS-302, one of two Ki-84s that were taken by U.S. forces from Utsunomiya to the U.S. for extensive flight-testing. In this recognition photograph, the plane is seen re-painted in a color scheme similar to what the IJA used on their planes in the field. During testing in early 1946, this aircraft, powered by 140-octane gasoline, was able to achieve the remarkable speed of 689km/h in level flight at combat weight while demonstrating better climbing and maneuverability than the P-51H and P-47N it was tested against. This led American testers to proclaim it the top Japanese fighter of the war.

▶テレビ映画出演のため、完全なオーバー・ホールを経て、17年ぶりに飛行可能状態に甦った1446号。塗装はふたたび全部落とされて、ジェラルミン地肌に戻った。胴体に記入された"N3385G"が、連邦航空局の認可をうけた飛行許可機であることを示している。垂直安定板の"AIR MUSEUM"は、本機のオーナーであるエド・マロニー氏が主宰する、オンタリオ航空博物館の所有機を示す。1963年6月末の撮影。

1446 takes to the skies for the first time in 18 years in this late June 1963 photo. The plane had been overhauled for TV-movie filming purposes. It appears here in an unpainted scheme. The 'N' number on the fuselage proves that the FAA has deemed the plane airworthy. Note as well the "Air Museum" marking on the tail, signifying the Ontario Air Museum operated by Mr. Ed Maloney.

パイロットたちは、"Ki-84は大戦中の日本戦闘機のベスト・ワン"という評価を下したのである。"大東亜決戦機"という華々しいキャッチ・フレーズに背き、日本陸軍から不本意な評価をうけた四式戦と中島飛行機(株)設計スタッフにとって、これはせめてもの救いだったろう。

1446号機に話をもどそう。1946年6月、1446号は米国国立航空博物館(のちのNASM)に移管され、保存されることになったが、1950年代に入って同博物館の運営予算が削減されたことにより、他の日、独機も含めて必要度の低いものを手離すことになった。1446号もそのなかに含まれ、1952年に、当時、カルフォルニア州のオンタリオ航空博物館長だったエド・マロニー氏が払い下げをうけることになった。

同博物館は、1446号に奇抜なマダラ迷彩をほどこし、映画などへの"出演"でギャラを稼いだが、1963年にはテレビ映画のなかで、米戦闘機と空中戦らしきものを演ずる必要から、飛行可能状態に整備することとなり、ロサンゼルスのギャレット・エアリサーチ航空部門の手によって、オーバーホールが行われた。そして、6月25日、同地のインターナショナル空港において、17年ぶりに飛行したのである。ちなみに、このとき本機を操縦したのは、第二次大戦中、ヨーロッパ戦線において、P-51マスタングに乗り、ドイツ機25機撃墜を果たしたエース、ウォーカー・マヒューリン氏であった。

その後、1446号は零戦、雷電とともに博物館の人気展示機として、多くの来館者を楽しませたが、1970年代に入ると同博物館(1970年にプレーンズ・オブ・フェイムに改称)の経営が悪化したため、1446号はエド。マロニー氏の友人の1人、エア・コーポレーション社長、ドン・ライキンス氏に移管されることになった。

しかし、ライキンス氏個人では1446号の維持は困難なため、同氏は日本へ売却することに決定、1973年に日本オーナー・パイロット協

as shoddy materials and workmanship in some components, especially the exhaust stacks.

It must be noted that during these tests, the plane had an advantage it never had during the actual war: high-grade gasoline. 87-octane was the best that Japanese forces could provide their planes during the conflict, and although "official" data on the Ki-84 indicated a top speed of 624 km/h, in practice the aircraft really could only reach 580-590km/h. However, with 140-octane gas in its tank, the normally troublesome Ha-45 was able to pull the Hayate through the skies at a top speed of 689km/h in level flight at 20,000ft., a stunning 60km/h+ more than the Japanese were ever able to produce, and just slightly faster than the P-51D-25-NA and P-47D-35-RA at the same altitude.

Following these tests, the U.S. evaluators proclaimed the plane the best Japanese fighter of the war, helping it to finally live up to its Japanese propaganda nickname of "The Great East Asian Fighter" and somewhat vindicating the design team and manufacturer from the Hayate's less-than-impressive battlefield performance.

1446 was eventually stored for the proposed forerunner of the National Air and Space Museum until 1952. However, budgetary considerations forced the museum to sell a number of their aircraft at this time, and 1446 was among these, being purchased by the now-famous aircraft collector Mr. Ed Maloney for his Ontario Air Museum. For a few years, the now-repainted 1446 was static displayed at the museum and used as a prop in a few movies (1954's *Never So Few*, et.al.). However, in 1963 the call came to film an aerial combat scene with the plane and it was decided to restore it to airworthy condition. Following an overhaul, 1446 took to the skies for the first time in 18 years on June 25. At the controls was no less than 25-kill WWII ace

◀1960年代後半から1970年頃にかけて、ふたたび上面ダークグリーン、下面グレイ塗装をほどこし、尾翼に飛行第102戦隊マークを描き込んで、オンタリオ航空博物館に展示されていた1446号。写真のように、ときとして野外に展示されたが、カルフォルニアの乾燥した気候と、手厚い保護のおかげで、機体コンディションは良好に維持されていた。

Following its media appearance, 1446 once again donned a green/gray scheme for display at the Ontario Air Museum in the late sixties to around 1970. As seen here, the plane was displayed outdoors from time to time, but in the dry California weather, and with careful maintenance, this didn't significantly harm its condition.

▶1973年10月2日、千葉県の陸上自衛隊木更津基地において催された、1446号の引き渡しセレモニー。カウリングや機首パネルを外して内部まで見せていたのが印象的だった。本機の前で椅子に腰掛ける人物は、右から購入者の後閑氏、ドン・ライキンス氏、エド・マロニー氏。

A scene from the hand-over ceremony at Kisarazu in Chiba, October 2, 1973. Note that the plane had its panels removed as well for complete viewing. Seated in front of the aircraft, from the right, are the buyer, Morinao Gokan, Don Lykins, and Ed Maloney.

◀上写真のセレモニー終了後、左後方から撮った1446号。尾翼の102戦隊マークは、方向舵羽布張り替えによって途切れてしまっている。このあと、本機は飛行してみせることになっていたが、ハ45エンジンが不調で始動できず、翌日に非公開で木更津上空を飛んだ。

1446 seen from the left rear after the conclusion of the ceremony. Note how the replacement of the rudder's fabric has erased half of the tail marking. At demonstration flight was planned for later in the day, but the Ha-45 wouldn't turn over and it was cancelled. A successful test flight was carried out the next day out of the public eye.

会会長を務める、後閑盛直氏が購入した。そして、同年9月、船便にて千葉県の木更津港に到着し、10月2日、ほど近い陸上自衛隊木更津基地において、エド・マロニー、ドン・ライキンス両氏も招待しての、引渡しセレモニーが行われた。

当日は、筆者もマスコミ関係者の一人として取材に赴き、初めて間近にみる実物の大戦機に、大いなる興奮を味わったことを、今でもはっきりと覚えている。予定では、午後に展示飛行が行われることになっていたのだが、あいにくハ45エンジンが不調で、キャンセルになってしまい、その勇姿を拝むことができなかった。

しかし、翌3日には非公開ながらテスト飛行に成功、10月7日から開催された入間航空ショーの最終日に、1446号はドン・ライキンス氏の操縦により、見事な飛行を披露、テレビ、新聞、雑誌などマスコミは、28年ぶりに故国の空を飛んだ旧大戦機として、大々的に報じた。

とにかく、完成してから30年近くもたった古典機、それもひと一倍気難しいエンジンを搭載している機体を、このように飛行可能状態に維持してきた、エド・マロニー、ドン・ライキンス氏をはじめとする、米国人の努力には、ひたすら頭の下がる思いがする。

Walker "Bud" Mahurin.

Following this flight, the aircraft was again displayed at the museum along with a Raiden and Zero. However, in the early 1970s, the now "Planes of Fame" Air Museum ran into financial difficulties, and 1446 was sold to Mr. Ed Lykins. Mr. Lykins quickly realized that it would be difficult to maintain 1446 on his own, and he decided to sell the plane to Japanese owners. In 1973, Mr. Morinao Gokan, president of the Japan Owner-Pilot Association, purchased the aircraft and had it shipped to Japan. It was brought in September of that year to the Kisarazu airbase of the JASDF in Chiba prefecture.

On October 2nd, the aircraft was showed off to the press and a demonstration flight was scheduled, but engine trouble scrapped those plans. Though not public, the aircraft made a short test flight the following day, and on the 7th, 1446 flew in front of thousands on the final day of the Iruma air show with Mr. Lykins at the controls. This was the first time the Japanese public had seen one of their own country's WWII aircraft fly in their home skies in 28 years, and the mass media gave it coverage

◀塗装をきれいに塗り直し、尾翼には新たに飛行第11戦隊第1中隊を示す雷光マークを描き、1973年10月5日から11日まで埼玉県入間飛行場にて開催された「国際宇宙ショー」で元気な飛行を披露した1446号。操縦者はもちろん、ドン・ライキンス氏。写真は、着陸態勢に入ったところで、特徴ある"蝶型フラップ"がいっぱいに下がっている。航空ショー期間中は、あいにく天候が悪く、1446号が飛行したのは、最終日だけだった。

With new paint and the markings of the 11th Sentai, 1st Chutai on the tail, 1446 struts its stuff for the crowds at Iruma on October 10, 1973. Don Lykins was at the controls of the plane for this flight. Note the fully-deployed "butterfly" combat flaps as he prepares to land. Bad weather during the show kept 1446 from flying except for this performance on the final day.

▶ハ45エンジンの轟音を響かせて、離陸した直後の1446号。木更津基地での引渡しから、入間航空ショー、そして保管場所に定められた、富士重工業・宇都宮南工場までのフェリーと、この間ずっと整備を担当してくれたのもマロニー氏のプレーン・オブ・フェイムのメカニック達であった。

1446 seen just after take-off on one of its few flights after returning to Japan. Ed Maloney's mechanics kept the plane in top condition during the test flight, air show and ferry flight to Utsunomiya where the plane was to be stored at Fuji Heavy Industries.

◀入間航空ショーの後、保管場所の富士重工業・宇都宮南工場にて、関係者向けの飛行を行った際のスナップ。油圧系統の不調のため、左主脚が充分収納されず、車輪カバーが開いたままになっている。結局、このときの飛行が、1446号にとってラスト・フライトとなり、以後二度と日本の空を飛ぶことはなかった。

A snapshot of 1446 taken during a private demonstration flight at Utsunomiya, after the arrival of the plane at Fuji Heavy Industries. Low hydraulic pressure appears to be preventing the plane from fully raising its left gear. This was the last time 1446 ever flew.

ただ、残念なことに、日本では古典機を良好なコンディションに保つことが、どれほど大変かを真に理解している人は少なかった。セスナ機をあつかうのとは訳がちがうのだ。整備の問題もそうだし、本機を操縦できるパイロットなど日本には1人もいない。別に、購入者の後閑氏を批判するわけではないが、1446号は、ドン・ライキンス氏らの手を離れた時点で、その飛行可能なコンディションを奪われてしまう運命にあった。

実際、入間で開催された航空ショーにおける飛行が、日本における1446号の最初で最後の晴れ舞台となったのだから……。

入間の航空ショー以後、1446号はかつてのメーカー中島飛行機(株)の後身である、富士重工業(株)宇都宮南工場の敷地内に野外保管され、静かな日々を送っていたのだが、所有者の後閑氏は死去したこともあって売却されることにな

appropriate to the significance of the event.

Ironically, this was to be 1446's public flight. The expertise and finances (not to mention a pilot!) to keep 1446 in airworthy condition simply did not exist in Japan at the time and the last remaining Ki-84 quickly fell into disrepair. The plane was displayed outside of Fuji Heavy Industries' (the post-war name of Nakajima) Utsunomiya South Plant for a few years, and following the death of Mr. Gokan was sold to the Arashiyama museum in Kyoto around 1980.

Most unfortunately, the private owners of this museum understood little about the value of the aircraft or how to display it properly. Seldom cleaned (reports are heard of spider webs inside the cowl), and subjected to rough handling by manner-less visitors who were not restricted from touching the aircraft, 1446 suffered through 10 difficult years

▲富士重工業・宇都宮南工場の敷地内で、野外保管されていた、1976年2月当時の1446号。排気管後方の汚れが、2年4ヵ月前の飛行を物語っている。この時点では、整備すればいくらでも飛行可能だった。

A photo of 1446 seen outside of Fuji Heavy Industries Utsunomiya South plant in February of 1976. The exhaust stains from its flight 2-1/2 years previous testify that this was a very much a living aircraft. At this point, a little maintenance would have been enough to get her back into the sky.

▲1989年8月、京都・嵐山美術館に展示されていた頃の1446号。保存、展示状態が悪く、機体の劣化が進んで、飛行不可能になってしまった。

Here's how 1446 looked at the Arashiyama museum in Kyoto in August 1989. With little or no attempt at preservation, the degradation of the aircraft's structures is quite apparent. The plane was clearly no longer airworthy at this point.

▲1997年3月から、知覧町の特攻平和記念館にて一般展示されている1446号。

1446 as it looked in March of 1997, on display at the Tokko Heiwa Kinen-kan, Chiran, Kagoshima.

り、1980年ころ、最終的に京都の嵐山美術館に引き取られることになった。

1989年夏、筆者は同館を取材に訪れ、木更津基地以来16年ぶりに、1446号に“再会”したのであるが、残念なことに目の前にある機体は、かつての“生きた実感”にはほど遠い、見る影もなく、各部に錆が浮き、操縦室にはクモの巣が張った、哀れな状態だった。

聞くところによれば、来館者のマナーの悪さも相当なもので、子供はおろか、大人までが主脚カバーに足を掛けたり、胴体点検パネルなどを剝がそうとしていたらしい。事実、本機のかたわらに展示してあった零戦六二型は、左主脚カバーの縁が内側に折れていた。情けない話である。購入者の勝手ではないかと反論されるだろうが、筆者はあえていいたい。現存大戦機の貴重な価値が少しでも理解できるのであれば、保存、展示法にもっと配慮があって然るべきではなかったかと……。

嵐山での10年以上におよぶ過酷な保存、展示により、1446号は飛行可能どころか、機体の原形維持にも黄信号が灯ってしまった。

幸いというべきか、1990年代に入り、嵐山美術館は閉館することとなり、1446号は暫定的に南紀・白浜に保管されたのち、鹿児島県・知覧町の特攻平和記念館に引き取られ、1997年3月、同館の増改築完成を機に一般展示さることになった。

エアコンの完備した室内、マナーの悪い来館者によって痛めつけられないような展示法とあいまって、どうやらこれ以上の機体の劣化はおさえられそうだ。ロールアウト以来、すでに54年が過ぎた1446号。もはやふたたび飛行することはかなわぬが、これ以上流転することなく、この特攻平和会館が、末長い安住の地にならんことを切に祈りたい。

Tokko Heiwa Kinen-kan, Chiran

知覧特攻平和記念館

住所	鹿児島県川辺郡知覧町郡 17881	Address:	17881 Kori, Chiran-machi, Kawanabe-gun, Kagoshima, Japan
電話	0993-83-2525	Tel:	0993-83-2525
開館時間	9:00～17:00(入室は16:30まで) 年中無休(ただし、都合により休館することがあります)	Hours:	Open daily 09:00 to 17:00

until the Arashiyama museum closed its doors in the early 1990s. Following the closing, there are reports that the plane's wings were severed in order to make storage easier.

In the mid-1990s, 1446 finally received new owners, this time in the form of the new Tokko Heiwa Kinen-kan located in Chiran, Kagoshima prefecture. The plane has been on display there since 1997 and is now in an air-conditioned hall and protected from the prying fingers of curious visitors. Though we will probably never see the plane take to the skies again, we can only pray that 1446 has finally found a home where its historical value will be faithfully preserved for generations to come.

機首 NOSE

1

2

写真3：取り外された着脱式パネルの裏側。疾風の着脱式パネルには格子状に補強がはいり、機体強度をそこなわないよう頑丈にできていた。手前のパネルはエンジンカウリング下面のもの。滑油冷却器フェアリング部があたる部分は切り欠かれている。

Photo 3: The interior side of the removed panels. They were well-reinforced to avoid creating any structural weakness when installed on the aircraft. The foreground panel is from the underside of the cowl. Note the cutout to clear the oil cooler fairing.

写真1-2：カウリングや点検口のパネルをはずした四式戦闘機"疾風"（キ84甲）の機首をみる。エンジンは中島ハ45（海軍名称「誉」）。本機が搭載したのは21型（ハ45-21）である。この空冷星形18気筒エンジンを搭載した疾風は、旧日本陸軍初の2000馬力級戦闘機であった。ハ45は世界に類を見ない2000馬力級小型高性能エンジンとして開発されたが、性能を極限まで追求した設計と構造から故障が多く、その結果、戦時中の工作精度の低下とあいまって搭載機の稼働率を低下させてしまった。

Photos 1-2: The Army Type 4 Fighter "Hayate" (Ki-84 Ko) seen with its cowling access panels removed, revealing the Nakajima Ha-45 "Homare" engine within. A model 21 (Ha-45-21) powered this particular example. The air-cooled, 18-cylinder Homare was the first 2,000hp engine put into use by the IJA, and remarkably compact for the power it generated. However, its design strained the limits of the technology of the day, and it suffered from severe fuel system problems. Oil loss was also troublesome and these factors seriously degraded the combat readiness of Hayate-equipped units.

3

4

5

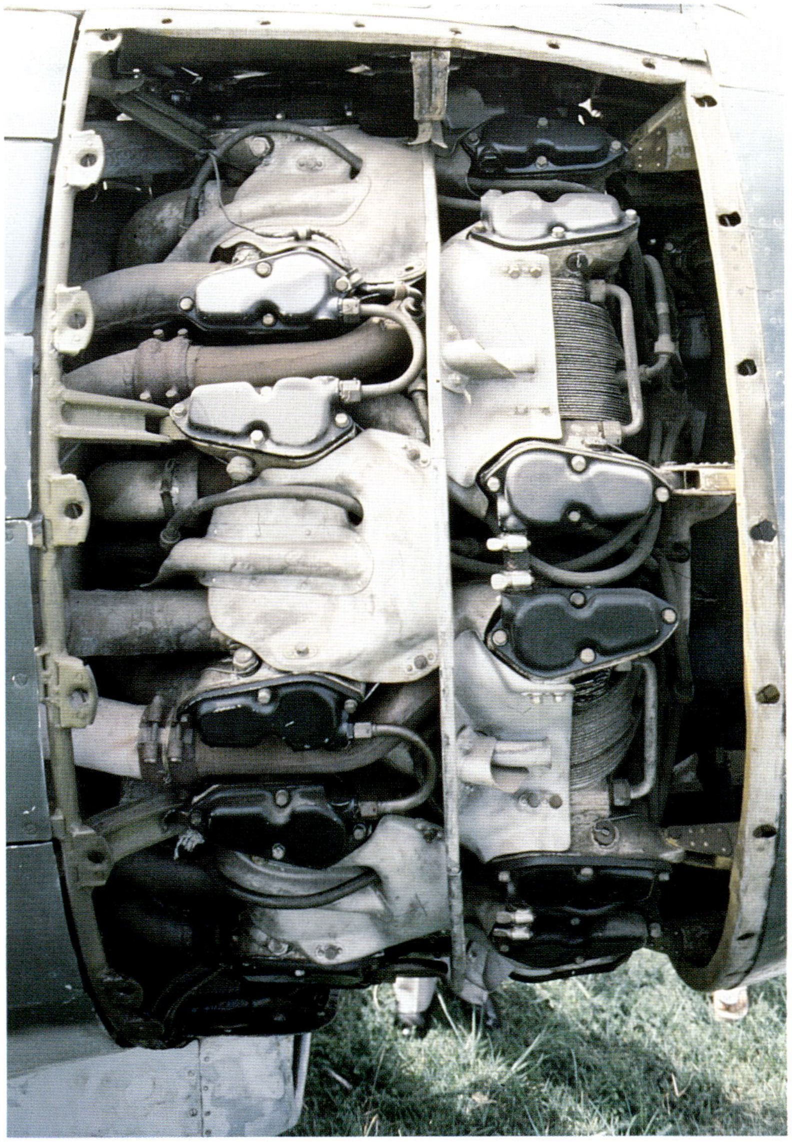

6

写真4：疾風のハ45-21エンジンを左斜め下から見上げたところ。1180㎜という小直径にまとめられたハ45エンジンの特徴をいかすため、疾風のカウリングは最大幅1260㎜という、2000馬力級エンジン搭載機としては類をみないコンパクトなものにおさえられた。写真からもほとんどエンジンの直径ぎりぎりにまでしぼられたカウリングのようすがわかる。

写真5：前列シリンダーのクローズアップ。プッシュロッド、コード類、異常に目の細かい冷却フィンに注目。2000馬力級エンジンをめざして開発されたハ45は、はじめ1800馬力エンジンの11型が採用され、冷却フィンの改良などに苦心した結果、2000馬力を達成した21型が誕生。これが疾風の搭載エンジンとなった。

写真6：機首左側からみたハ45-21エンジン。エンジンはカウリング内の限られた空間を占め、シリンダーの間隔も狭い。さらにそのすきまへ、後列シリンダーに冷却空気を導くためのバッフルプレートや、シリンダーのあいだを縫うように取り付けられた排気管、配線、パイプ類などが組み込まれており、構成はかなり複雑である。

Photo 4: The Homare 45-21 seen looking up from the lower left side. In order to fully exploit the small, 1180mm diameter of the engine, the Hayate's cowl was also designed as compactly as possible at just 1260mm diameter. The photos show well just how little space was left over inside the nose of the plane!

Photo 5: A close-up of the forward row of cylinders, showing well the push rods, oil lines, cooling fins and other details. The first version of the Homare, the model 11, came in at 1,800hp. Modifications to the cooling fins and other details were ultimately able to produce the sought-after 2,000hp in the model 21. It was this model which was installed in the Hayate.

Photo 6: The engine seen from the left side of the nose. Not only is there little space left between the engine and the cowl, but the space between the cylinders is also extremely tight. This required a complicated design to snake the fluid lines and exhaust pipes between the cylinders, and to install air baffles to channel cooling air back to the rear row of cylinders.

機首 NOSE

7

8

9

写真7：疾風の機首を正面からみる。疾風に装備されたプロペラは「ペ三二」全金属製4翅電気式定速プロペラ。フランスのラチエ社が開発中だったプロペラの製造権を購入し、日本独自の改良を加えたもの。重量はスピナー込みで185kg。カウリング内のスピナー上方にみえる棒状のものは、マニュアルでプロペラピッチを微調整するための操作ロッド。

写真8：カウリング正面上部の気化器空気取り入れ口を上からみる。プロペラピッチ微調整用ロッドはピッチの調速機に連結されている。この電気式調速機でプロペラピッチを制御し、変更を自動的に行う。

写真9：左後方から機首の機関砲発射口弾道溝をみる。増加試作機の後期型と生産型の砲口覆いは写真のように短い。キ84甲の胴体武装であるホ103 12.7㎜機関砲は、この展示機では取りはずされている。ちなみに、対爆撃機用として生産されたキ84乙の胴体武装はホ五 20㎜機関砲であった。

写真10：左斜め前からみた疾風の機首。

Photo 7: A view of the plane's nose from directly ahead. The Hayate mounted the "Pe-32" propeller, a four-blade, fixed-speed, variable-pitch electric prop that was based on a design under development by Ratier of France. The Japanese purchased the rights to this design and further refined it domestically.

Photo 8: The carburetor intake on the upper side of the nose seen from slightly above. The small rod seen inside the cowling just below it is the propeller pitch control rod and is connected to the pitch control itself that automatically controlled pitch electrically.

Photo 9: A view of the nose machine gun muzzle cover on the left side of the upper cowl, seen from the rear left. The Ho-103 12.7mm MG was standard nose armament on the Hayate, but this example does not have them installed. Second-batch supplementary prototypes and production aircraft had short muzzle covers, as seen here.

Photo 10: An overall view of the nose of the plane from the forward left.

10

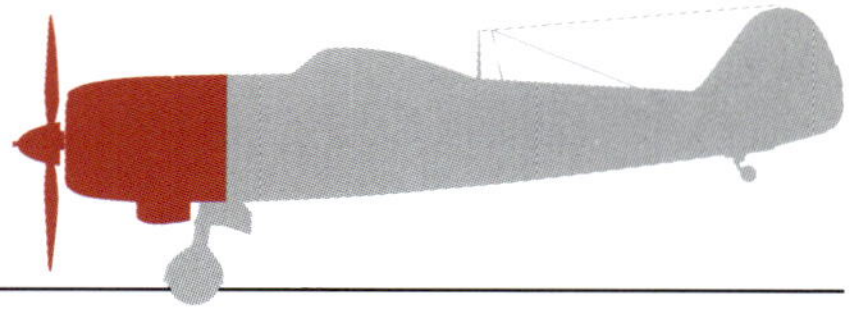

11

写真11-13：カウリングの開口部からみたハ45エンジン正面。プッシュロッドやコード類の構成に注目。シリンダー同士の間隔がせまいのが、この写真からもわかる。スピナー上部にみえるプロペラピッチ調速機のディテールにも注目。この調速機はプロペラ回転数に同調するセンサー部と、ピッチ変更モーターを動かすためのスイッチを内蔵している。操作ロッド下方にみえるコードは、モーターに電力をおくるためのもの。なお、写真ではレンズの関係でカウリング前面の開口部が広くみえるが、実際にはかなり絞り込まれている。カウリング前面開口部の左右幅は80cm。カウリング上部の気化器空気取り入れ口中央には縦に仕切り板がある。また、取り入れ口の奥に、気化器空気取り入れ導管(ダクト)前部の先端がわずかにのぞいているのが写真から確認できる。

Photos 11-13: Views of the Ha-45-21 engine from the front. Note again the cramped conditions between cylinders intermingled with the push rods and fluid lines. The propeller pitch control is also shown well, just above the spinner. The unit contained a sensor that detected propeller rotation speed and adjusted pitch accordingly. Note the wire leading from the bottom of the unit, which carried power to the pitch control motor itself. This photo was taken with a wide-angle lens, making the cowl opening seem somewhat larger than it actually was, a mere 80cm across! Note the small divider in the middle of the carburetor intake. The carburetor air duct is just barely visible behind the opening.

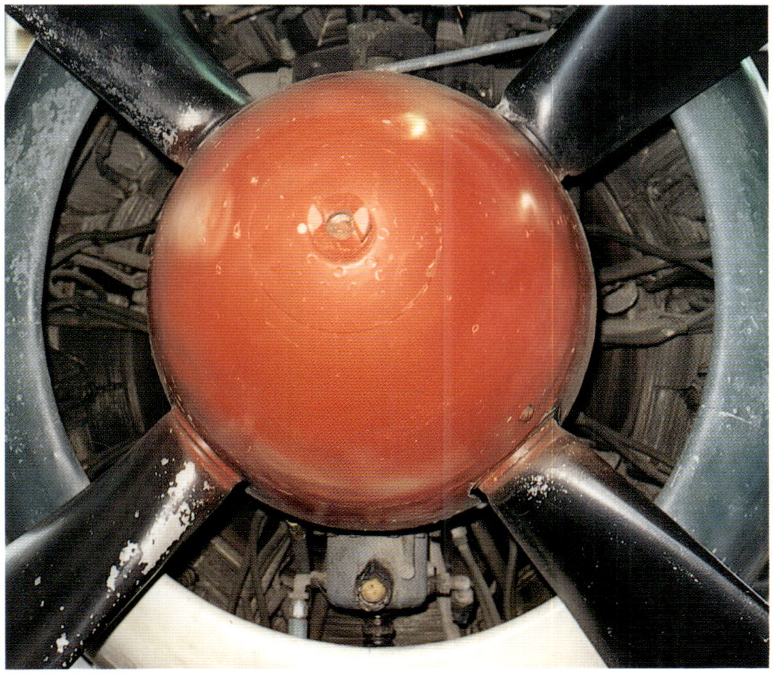

12

13

機首 NOSE

14

15

写真14-15：疾風の機首右側面。写真左側にみえるタテに並んだスリット状の部分はエンジンルームの通風をよくするためのもの。この直後に胴体第1フレーム(円框)がある。

Photos 14-15: A view of the right side of the nose. The vertically-arranged slits visible at the left side of the photo just above the leading edge of the wing were to help improve the passage of air through the cowl. The first fuselage frame was located directly behind these.

16

17

18

写真16-18：機首右側のカウルフラップと排気管をみる。上部排気管後方にみえる楕円形をした2カ所のパッチは、ホー〇三機関砲用発射ガス抜き穴を塞いだもの。写真18は排気管を後方からみたところ。金具で胴体に止められている排気管内側のディテールに注目。

Photos 16-18: Views of the cowl flaps and exhaust pipes on the right side of the nose. The round patch visible just behind the upper exhaust pipe covers the exhaust gas port for the now-absent Ho-103 nose armament. Photo 18 is a view of the exhaust pipes from the rear.

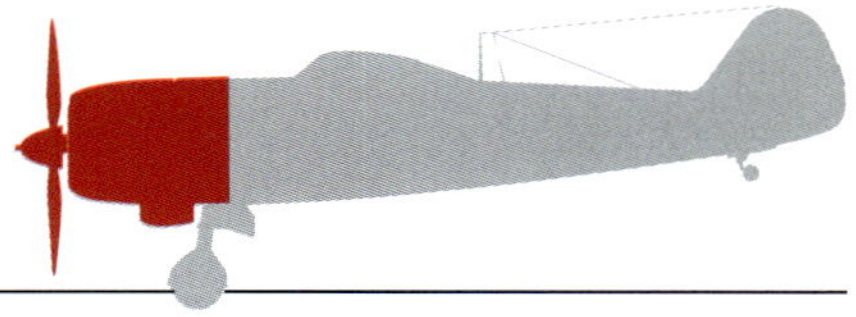

19

22

20

21

23

写真19：機首左側面。機首上面のホ－○三機関砲発射口と弾道溝が確認できる。こちら側の機関砲用ガス抜き穴もパッチで塞がれている。

写真20-21：写真20は左側後方からみた上部排気管のディテール。左側上部の排気管は2本に分かれているが、右側上部のものは写真21のようにシリンダー2本分をまとめて1本となっている。胴体に固定された排気管に注目。

写真22：機首左側のカウルフラップと排気管のディテール。中央にみえる4本の排気管のうち下の1本はオリジナルでないため形状が異なる。またその下方の排気管はシリンダ2つ分を1本にまとめてある。

写真23：後方からみた機首左側のカウルフラップと排気管。

Photo 19: The left side of the nose, showing well the nose armament muzzle covers and bullet channels. The gun exhaust has been patched on this side of the plane as well.

Photos 20-21: Photo 20 is a close-up of the upper exhaust pipe on the left side of the nose, actually two pipes side-by-side. As photo 21 reveals, however, the upper exhaust pipe on the right side combines two cylinders' exhaust into one pipe (see also photos on p.12).

Photo 22: A revealing shot of the cowl flaps and exhaust pipes on the left side. Note that the bottom exhaust pipe in the middle cluster of four is not original equipment, causing it to differ in shape from the others.

Photo 23: A view of the left-side cowl flaps and exhaust pipes taken from the rear.

機首 NOSE

24

写真24 :エンジンから防火壁まわり、および主翼のおもなパネルをあけたところ。防火壁後方にある縦長の開口部はホ一〇三胴体機関砲の弾倉収容部。残念ながら弾倉本体は残されていない。エンジン取り付け架と潤滑油タンクは黄色に塗られている。潤滑油タンクの容量は75リッター。エンジン取り付け架は胴体の第1フレームに組み付けられている。第1フレームは防火壁を兼ね、この部分を境に前方が機首、後方が胴体となる。

Photo 24: A view of the aircraft from the left, with nearly all important access panels removed. The smaller, vertical opening behind the firewall is to allow access to the nose armament's magazines, but unfortunately they are no longer present on this plane. The engine mount and oil tank, which held 75 liters, are painted yellow. The engine mount was secured to the first fuselage frame, which also doubled as the firewall and established a clear division between the plane's nose and fuselage.

写真25 :機体側面前部を左主翼の上方からみる。主翼上面のホ五20㎜機関砲用点検パネルも開けられている。エンジン取り付け架の上方にスペースがあいているのは、本来ここをホ一〇三機関砲の銃身が通るため。ホ一〇三機関砲用弾倉収容部後方は燃料注入口。外蓋が開けられ、赤く塗られた内蓋がみえる。また、その後方に飛び出した四角い棒はとび出し式の乗降用足掛け。

Photo 25: The opened-up aircraft seen from a vantage point above the wing. Note that wing access panel for the Ho-5 20mm cannon is also open. Note the relatively open area above the engine mount. Normally, the Ho-103 machine gun breeches would protrude into this area of the nose. The round opening behind the machine gun magazine access is a fuel fill. The cover is open, revealing the red paint within. The small protrusion behind and below this fuel fill is a retractable footstep for boarding the aircraft.

25

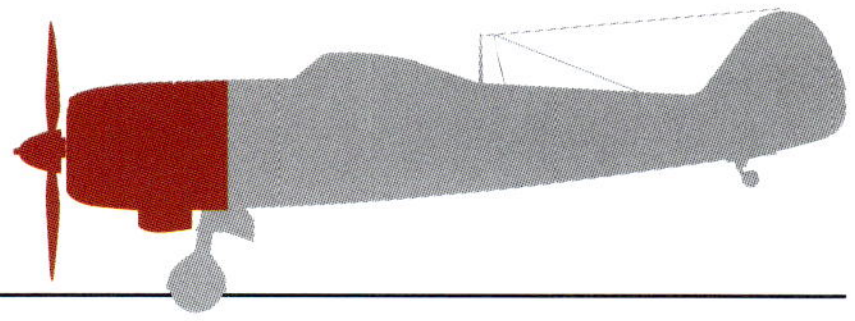

写真26-27：エンジン後方と取付架の構成。カウリング内部後方にのぞくダクトが気化器空気取り入れ導管後部。湾曲部の内部にも整流板が設けられている。ハ45はこの導管のかたちをわずかに変えただけで、最高速度が低下してしまうほど、デリケートなエンジンであったという。滑油タンクのまわりにはエンジン管制用のロッド、ワイヤーやコード類が張り巡らされている。写真27で防火壁後方、胴体上面のパネル開口部にみえるのは油圧系統用作動油タンク。本来この後方にあるはずの、メタノールタンクおよびその注入口は確認できない。復元機とはいえ、エンジンマウントまわりの構造が確認できるのは貴重。機関砲用爆発筒などのオリジナルディテールは63頁を参照。

Photos 26-27: Views of the engine mount and other structures immediately behind the engine, ahead of the firewall. The rear portion of the carburetor air duct is visible just behind the cowling. Though not visible in the photo, there were specially-designed airflow baffles present in this part of the duct as well. It is said that even a slight change in the shape of these would lower the top speed of the aircraft due to performance changes. The engine control rods and wires are clustered near the oil tank. The oil fill is visible in photo 27, just behind the firewall. There should be a methanol tank fill just behind that, but it's missing in this aircraft.

26

27

機首 NOSE

28

29

30

写真28：スピナーを取り外した機首の正面。疾風の「ペ三二」全金属製4翅電気式定速プロペラの直径は3m。2000馬力級戦闘機のプロペラとしては小さいといえるが、これは主脚を短くおさえて重量軽減をはかった結果、ロード・クリアランスの問題で直径が制限されたため。プロペラ回転軸先端にみえるのがピッチ変更用モーター。

写真29-30：取り外したスピナーのディテールをみる。写真29ではスピナーの止めネジが確認できる。

写真31：プロペラ回転軸先端にとりつけられたピッチ変更用モーター。このモーターでプロペラピッチ変更用の動力を供給する。モーターはプロペラ本体といっしょに回転し、定格は90W、直流。疾風の電気式定速プロペラは構造が複雑なことからトラブルが多く、エンジンの不調と共に実戦部隊で大きな問題となった。

Photo 28: The front of the aircraft, seen with the spinner removed. The all-metal Pe-32 four-blade electric fixed-speed propeller had a diameter of 3m. That seems small for a 2,000hp aircraft, but was necessary for ground-clearance reasons, as the Hayate had relatively short legs as a weight-saving feature. The pitch control motor is visible in the center of the propeller.

Photos 29-30: Close-ups of the removed spinner.

Photo 31: The pitch control motor in the center of the propeller. It rotates with the propeller. The motor was a 90W, direct-current device. This relatively complicated device caused maintenance problems in the field and, along with the engine, contributed to poor readiness ratios among Hayate-equipped units.

31

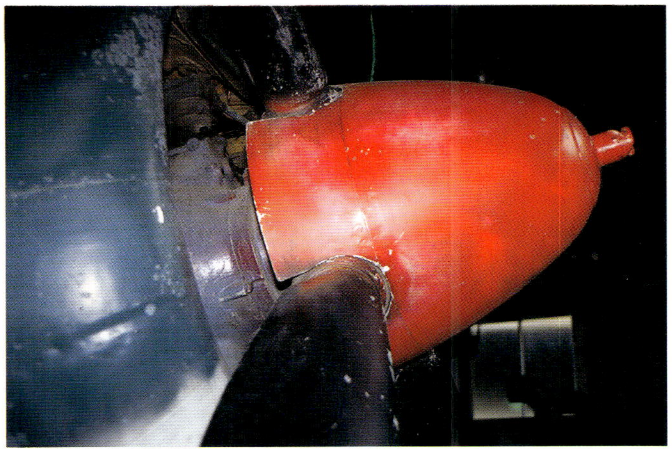

32

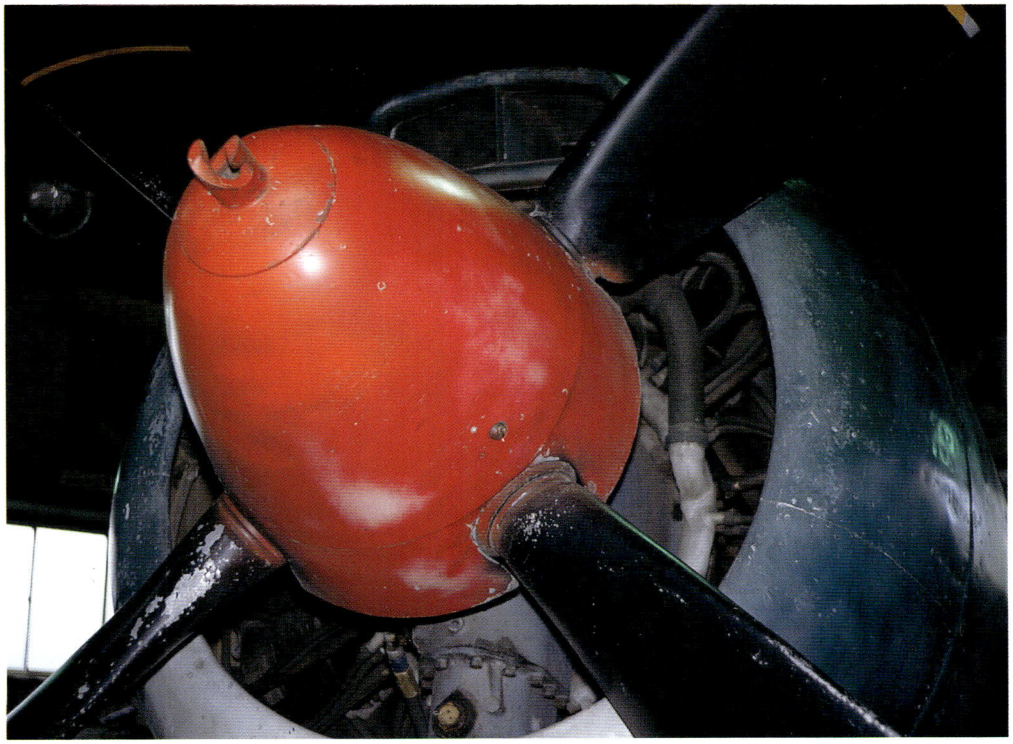

33

写真32-34：スピナーまわりの細部をみる。スピナーの直径は50cm。写真34でプロペラ軸の上にみえる装置が、プロペラと共に回転するピッチ変更用モーターへ電力を供給するカーボンブラシの保持器。

写真35-36：「ペ三二」のプロペラブレード側面と正面。

Photos 32-34: Close-ups of the area around the spinner. The spinner was 50cm in diameter. The small device visible above the propeller axle in photo 34 is the carbon brush assembly, which transmits electric power to the pitch control motor, which rotates with the propeller.

Photos 35-36: Views of the Pe-32 propeller blades.

34

35

36

機首 NOSE

写真37：疾風のハ45エンジンと滑油冷却器を見上げる。エンジンはカウリング内左右いっぱいにおさめられていることが、この写真からもわかり、後列シリンダーの冷却に問題があったこともうなずける。撮影したレンズの関係から、写真では滑油冷却器の大きさが実際より強調されている。

写真38-39：滑油冷却機を正面と右斜め前からみる。冷却器は機首下面へかなり張り出している。

Photo 37: A view of the Ha-45 and its oil cooler from underneath the aircraft. This photo also shows off well how the engine was packed into the cowling with little room to spare, and why cooling the back row of cylinders was difficult. Note that this photo was taken with a wide-angle lens, making the oil cooler seem larger than normal.

Photos 38-39: Views from directly ahead and the forward right of the oil cooler. The cooler protruded quite a bit below the nose of the plane.

37

38

39

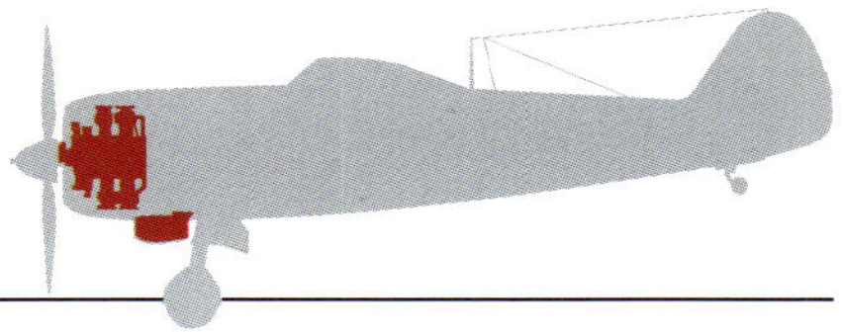

40

写真40 ：右後方からみた滑油冷却器まわり。温度調整用のシャッターは閉まっている。下部排気管のまわりにカウルフラップ開閉用ロッドの一部分が確認できることにも注目。

Photo 40: The oil cooler seen from the right rear. The temperature control shutter is seen closed. Note that some of the control rods for the cowl flaps can be seen near the lower exhaust pipes.

41

42

写真41 ：フェアリングにおさめられた滑油冷却器本体を前方からみる。冷却器本体は楕円形で、蜂の巣状の構造をした本体を通る空気が内部の環流管にあたり、潤滑油を冷却する仕組み。前方の張り出した冷却器をクリアするため、カウリングの一部は切り欠かれていることに注目（写真3参照）。

写真42 ：滑油冷却器左側面。フェアリングは前部、中央部と後部に3分割され、着脱可能。胴体へはボルトで固定される。

Photo 41: The entire cooler assembly from the front. Air would pass through these honeycomb openings to cool the oil pumped through pipes behind this. As mentioned above, note the cutout in the removable lower cowl access panel to clear the cooler.

Photo 42: The cooler seen from the left side. Note how the fairing is broken into three sections. These were bolted to the aircraft but could be removed for maintenance.

キャノピー&コクピット　CANOPY & COCKPIT

43

写真43-46：前部固定風防のディテール。風防直前に設けられた外気取り入れ口は開度を3段階に調節可能。写真44で後部フレーム上端の幅が太くなっているようにみえるのは、撮影レンズのゆがみのため。実際は写真45のように上下共同じ幅である。また、前部固定風防の前面には、本来厚さ70㎜の防弾ガラスがつくが、この復元機では取り外されている。写真46で風防内側に取り付けられたジャイロ・コンパスはもちろん復元後に米国内での飛行のため装備されたもの。本機が戦後も民間機として"飛行していた"ことのあかしである。

Photos 43-46: Close-up views of the Hayate's windscreen. The small air intake visible on the leading edge of the windscreen in photo 43 could be opened to three different sizes. Note that the rear edge of the windscreen frame appears larger at the top due to the wide-angle lens used to take the shot. The frame is consistent in width, as shown by photo 45. Normally, a piece of 70mm bulletproof glass would be attached to the front of the windscreen, but it has been removed from this example. The compass visible mounted inside the windscreen in photo 46 is not original equipment.

44

45

46

47

48

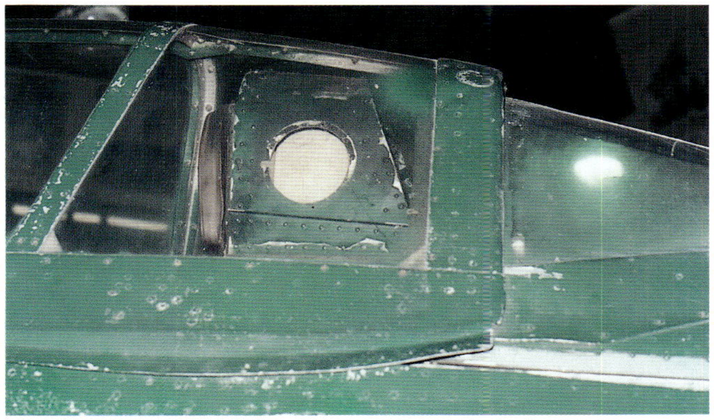

49

写真47：左側面から風防全体をみる。疾風の風防は前部固定風防、中央可動風防、後部固定風防の3分割式。

写真48：風防全開状態。中央可動風防は緊急時に風圧で飛散できる構造になっていた。

写真49：座席後方の転覆時保護支柱。取り付けられているパイロット用の頭当ては、オリジナルと異なる。支柱の丸い穴は主脚／尾輪用警報ブザー取り付け用。本機のブザーは欠落している。

Photo 47: A view of the entire canopy from the left side. The canopy was a very orthodox construction, with fixed front and rear sections, and a sliding middle portion.

Photo 48: The canopy seen fully opened. This could be jettisoned from the aircraft in an emergency.

Photo 49: A view of the roll bar in the rear of the cockpit, complete with a cushion for the pilot's head. The round holes in the bar are for installation of the landing gear warning buzzers, which are missing from this plane.

50

写真50-51：左斜め後方からみた疾風の風防。展示機の乗降用手掛けはバネ仕掛けの蓋が欠落し、ガイドレール後方にもオリジナルにはない補強が取り付けられている。

写真52：疾風の風防を後方からみる。写真から、本機の可動風防上部フレームの幅が中央前後で異なることを確認できる。これが復元時の都合なのか、戦時下の生産ラインで生じた何らかの理由によるものかは不明。

Photos 50-51: The canopy seen from the left rear. Note that the handhold below the cockpit has lost its spring-loaded cover, and that a non-original support of some kind has been installed immediately to the rear of the canopy slide rail.

Photo 52: Views of the canopy from the rear. The photo shows well how the upper frame on the moveable portion of the canopy varies in width near the center. It's uncertain whether this was due to some manufacturing factor, or whether this occurred when the plane was restored.

51

52

キャノピー&コクピット CANOPY & COCKPIT

53

54

55

写真53：復元機のコクピット内部をみる。疾風の計器板は上段に航法関係、中／下段にエンジンや燃料関係の計器を配するアレンジであった。残念ながら写真の復元機は計器の多くをアメリカのものに交換されている。ただし一番下の一列は、ほぼオリジナルのまま。計器板右下にみえる配電盤のパネル、フットバーもオリジナルとは異なるものが取り付けられている。

写真54-55：コクピット内部左側。前部固定風防下のハンドルは中央風防開閉用。写真55はエンジン操作関係のレバー群と、その後方にある昇降舵トリムタブ操作ハンドル。昇降舵トリムタブは写真151-155を参照。

Photo 53: The cockpit of the Hayate. The main instrument panel of the plane was arranged with navigational instruments in the top row of the panel, and engine and fuel related indicators in the middle and bottom rows. Unfortunately, this aircraft has had many of its instruments converted to American models. The bottommost row, however, is generally intact. The electrical box visible to the lower right of the panel, as well as the foot bar, are not original equipment.

Photos 54-55: The left side of the cockpit. The crank handle visible below the windscreen is to operate the sliding canopy. Photo 55 offers a close-up of the throttle quadrant as well as the trim control and other flight controls.

56

57

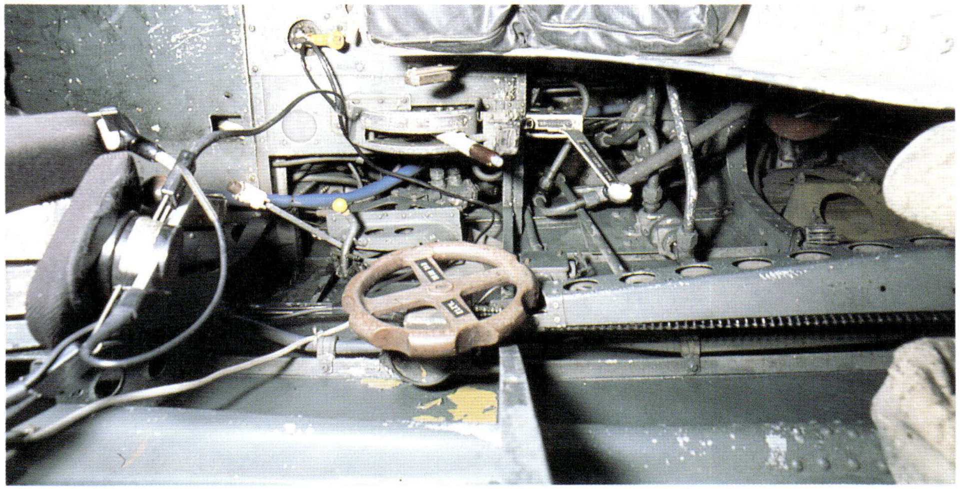

58

59

60

写真56：復元機のコクピット内部を上方から見下ろす。

写真57：復元機のコクピット内部右側。配電盤斜め上のレバー類は滑油冷却器操作レバーとカウルフラップ操作レバー（黄色）。その上の黒いボックスは日本製のオリジナルであり、写真銃用の計数器と思われる。配電盤手前のレバーは燃料管制装置。座席右横の丸い握りのついたレバーが、手動燃料ポンプである。

写真58：座席左横のレバー類。日本語で15←→閉←→30という表示が確認できるのはフラップ操作レバー。

写真59-60：座席、頭当てと転覆保護支柱のディテールをみる。頭当てはオリジナルではない。ちなみに疾風のコクピットには暖房設備がなく、電熱服で代用した。しかしこの電熱服ではからだ全体を暖めることはできず、さらに、すぐショートするなどの問題があったという。

Photo 56: A view down into the cockpit from directly overhead.

Photo 57: The right side of the restored cockpit. The levers diagonally above and to the rear of the non-original electrical box (lower right of the instrument panel) are to control the oil cooler and cowl flaps (the yellow lever).

Photo 58: A look down at the controls on the left side of the cockpit. Inboard of the red trim control wheel is a lever with the markings "15←→閉←→30" (the middle character is read "he" and means closed). This is the control for the aircraft's "butterfly" combat flaps.

Photos 59-60: Close-ups of the pilot's seat and the roll bar/headrest assembly. The headrest is not original. The Hayate had no cockpit heater. Pilots wore an electrically-heated flight suit instead. It's said that this garment didn't warm the entire body very well however, and frequently shorted-out!

胴体 FUSELAGE

61

写真61：疾風の機首から胴体前方にかけてのラインを左後方からみる。防火壁部分後方のパネルはホー〇三胴体内機関砲弾倉交換用の着脱パネル。疾風の前部胴体は零戦などと同様に組み立て後主翼と一体となる構造を採用していた。

Photo 61: The forward portion of the left side of the fuselage. The vertically-oblong panel just behind the firewall was to provide access to the Ho-103 nose-mounted machine guns' magazines.

写真62：胴体左側前方を真横からみる。胴体は第3風防後部にある、この写真では日の丸直前にみえるライン（第9フレーム）で前後に分割される。

写真63：ホー〇三胴体内機関砲弾倉交換用の着脱パネル、クローズアップ。この着脱パネルにある2カ所のロックはいわゆるワンタッチ式。写真ではツメが起きている。

写真64：疾風の胴体平面形は機首から尾端までほぼ直線的に結ぶラインが採用され、隼や鍾馗のようにカウリング直後から急激に絞るものとは異なる。カウリング後方から胴体へかけてのラインがなだらかなことが、この写真からもわかる。

Photo 62: An overall view of the left side of the fuselage. The fuselage splits into two at the 9th frame, located immediately behind the rear canopy section, or just forward of the fuselage *hinomaru* marking in the photo.

Photo 63: A close-up of the Ho-103 magazine access panel. The panel was secured in place with two one-touch twist locks. These are easily identifiable in the photo with their tabs seen in the raised position. These would fold flat for flight.

Photo 64: A view of the mid-section of the fuselage seen from the rear. The fuselage cross-section of the Hayate narrows from the nose to the tail in a straight-line design. This differs from the Hayabusa, Shoki and some other designs that quickly narrow immediately behind the cowl.

63

62

64

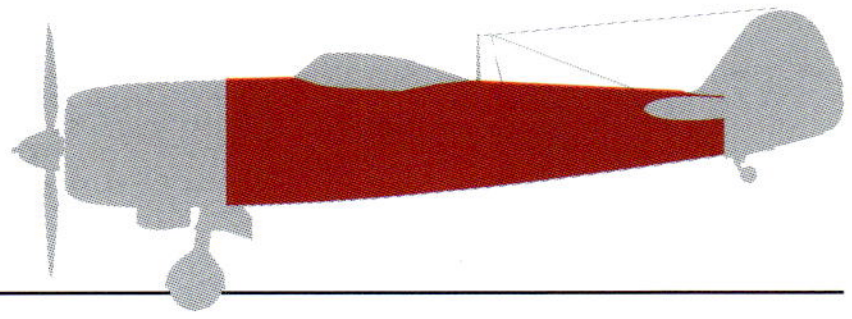

65

66

67

68

69

写真65-66：胴体左側面の乗降用足掛け。足掛け真上のボタンを押すと写真66のように飛び出す仕掛け。

写真67：胴体左側面の乗降用手掛け。通常はスプリング式の蓋が閉じているが、蓋を押すと手掛けになる。

写真68：胴体左側、フィレット下面の引き込み式足掛け。

写真69：胴体左側面にある燃料注入口のクローズアップ。

Photos 65-66: The boarding footstep on the left side of the fuselage. Pushing the button above the step causes it to spring out of the fuselage, as seen in photo 66.

Photo 67: A close-up of the hand hold on the left side. This is covered with a spring-loaded panel that is pushed in when used, as seen here.

Photo 68: A view of the retracting footstep under the wing root fillet that deploys for boarding.

Photo 69: A close-up of the fuel fill on the left side of the fuselage.

胴体　FUSELAGE

70

71

写真70、72：後部胴体左側にある点検作業口。左側のみにある。

写真71：点検作業口カバーの裏側。疾風では着脱パネルの多くにいわゆるワンタッチ式ロックを採用し、各種作業の効率化をはかっていた。写真からロックの構造が確認できる。

写真73：点検作業口カバー。展示上の都合からだろう、カバーはテープで塞がれ、その上から塗料を塗られてしまった状態。

Photos 70, 72: The fuselage interior access hatch. This is present only on the left side of the aircraft.

Photo 71: A view of the interior of the removed hatch. Unlike many earlier IJA and IJN aircraft, the Hayate used many of the so-called "one-touch" fasteners on most of its access panels to improve maintenance efficiency.

Photo 73: The hatch before removal. Note the tape that was used to help secure it. This is, of course, not original!

72

73

74

75

写真74-75：後部胴体内部からコクピット方向をみる。座席後部のディテールがわかる。本機の座席支柱には本来取り付けられているはずの緩衝ゴムがなく、座席は支柱に直接取り付けられているようにみえる。機体内部のジンクイエローはアメリカで塗装されたもの。

写真76：点検作業口から胴体内部をみる。内部に張りめぐらされたワイヤーは方向舵と昇降舵の操作用。

写真77：後部胴体の内部。胴体はフレームとストリンガー（縦通材）を外板で覆ったセミモノコック構造。

Photos 74-75: The inside of the fuselage, looking from the rear towards the cockpit. The pilot's seat should be connected to the support posts with shock-absorbing rubber connectors, but it appears to now connected directly to the posts. The yellow color seen here is zinc-yellow that was painted when the aircraft was in the United States and is not original.

Photo 76: The fuselage interior farther back, close to the access hatch itself. Note the control cables leading back to the rudder and elevator.

Photo 77: Looking straight back towards the tail. The standard, semi-monocoque frame and stringer construction is quite evident in this view.

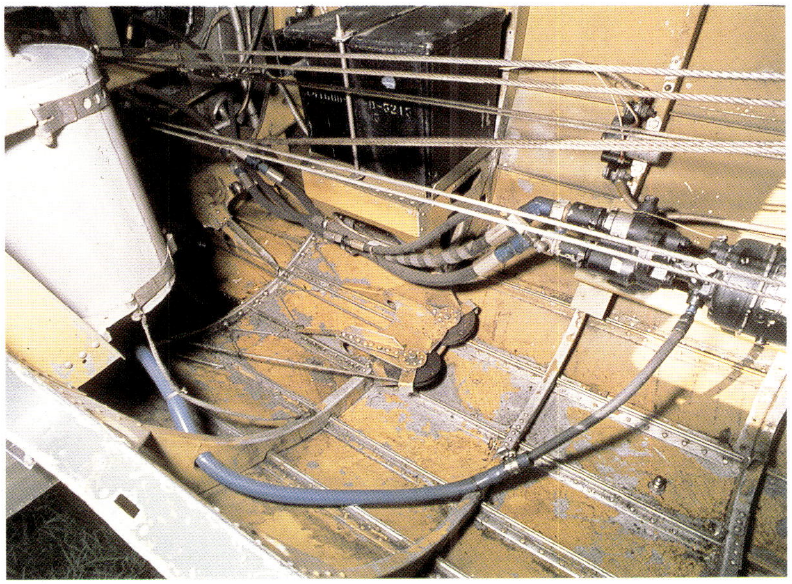

76

77

78

写真78：後部胴体左側の点検作業口付近をみる。点検口は第10フレームと第12フレームのあいだに開けられている。

Photo 78: A view of the area around the access panel. A comparison with photo 80 confirms the lack of such a panel on the right.

79

80

写真79-80：後部胴体右側。前部胴体と主翼を一体にする構造は、中島が九七式戦闘機で採用した方式。胴体と主翼を分割式とするより結合部の構造重量を軽減できる利点がある。胴体の前後は6㎜径のボルト47本、5㎜のボルト8本に、それぞれ同数のナットで接合した。

Photos 79-80: Views of the right side of the fuselage. Nakajima began using the one-piece wing and forward fuselage design with their Type 97 Fighter. Compared to a separate wing and fuselage assembly, this design lowered the weight of the required attachment hardware. The forward and rear fuselage were held together with 47 6mm bolts and eight 5mm bolts, as well as the corresponding nuts.

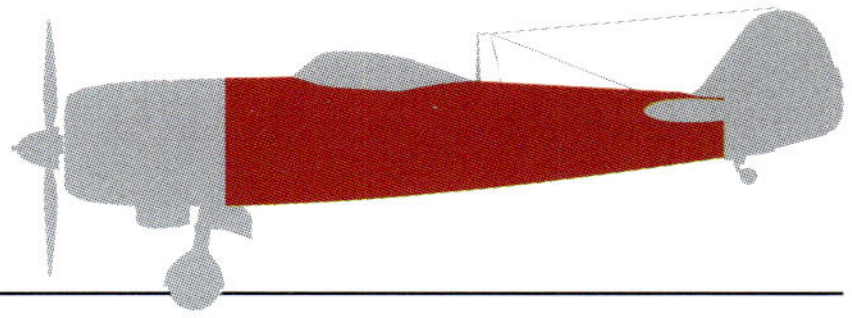

81

82

83

写真81-82：疾風の後部胴体。外観からではわからないが、胴体は九七式戦闘機以来の前後分割式を踏襲したうえで、さらに量産性を考慮した設計になっていた。上面のアンテナは民間飛行用の装備で、オリジナルにはない。

写真83：ほぼ真後ろからみた胴体。尾部に向かう平面形が直線的なラインで構成されていることが、この写真からもみてとれる。また、リベットのラインもよくわかる。胴体外板の重なり具合にも注目。胴体後部の外板は厚さ0.6mm。

Photos 81-82: Views of the rear portion of the Hayate's fuselage. Though not apparent from the photos, the Hayate's fuselage was of the front/rear split design used by Nakajima beginning with the Type 97 fighter and incorporated other refinements to aid mass production as well. The antenna on the upper side is not original equipment.

Photos 83: The fuselage seen from almost directly behind. The straight-line taper of the fuselage is clear in this view, as are the fuselage panel seams.

84

85

写真84-85 ：左主翼上面のホ五 20㎜機関砲点検ハッチを開けて機関砲収容部をみる（機関砲は未装備）。その外側にある桁方向にそって細長い長方形のハッチは、ベルト弾倉用パネル。裏返して置かれた長方形のハッチは、胴体武装のホ一〇三 12.7㎜機関砲用弾倉収納部の着脱パネル。また、その横のパネルが作動油タンク注入口のパネル。主翼燃料タンク注入口の外蓋も開いていることに注目。

写真86 ：アメリカ国立航空宇宙博物館（NASM）で、M2ブローニング機関銃（上）と共に展示されているホ五 20㎜機関砲。ホ五と疾風の胴体武装であるホ一〇三 12.7㎜機関砲は、このブローニング系機関銃をコピーしたもの。なお、ホ五は開発当時、量産用図面を確定しないままに生産を急いだことで、各製造元で異なったトラブルや故障を生じ、整備、運用上の混乱をおこしている。ちなみに旧日本陸軍は12.7㎜以上を機関砲、海軍は20㎜までを機関銃と呼称した。

写真84-85 ：同じくNASMに展示されている旧日本陸・海軍の機関銃弾と機関砲弾。上から海軍九九式1号20㎜機関銃用（プレートに九七式とあるのは誤り）、疾風が搭載したホ五機関砲用、3番目と4番目が海軍三式12.7㎜機関銃用、5番目が疾風の胴体武装であったホ一〇三機関砲用。

Photos 84-85: Views of the upper surface of the left wing with the access panels for the Ho-5 20mm cannon opened up (the cannon itself is missing). Though closed, outboard of the cannon access can be seen the oblong hatch covering the compartment for the cannon's magazine. Next to this, two removed panels are visibly lying on the wing. The upside-down one is the cover for the fuselage Ho-103 12.7mm MG magazine access, while the roughly square one covers the oil fill. The wing fuel fill cover is also open in this photo.

Photo 86: On display at the National Air and Space Museum (NASM) in Washington D.C. are an M2 Browning machine gun (above), and a Ho-5 20mm cannon. Along with the Ho-103 12.7mm MGs in the fuselage, the Ho-5 was a copy of the M2, though scaled up.

Photo 87: The display of IJA and IJN aircraft ammunition at the NASM. From the top are shells for the Navy Type 99 Mk.I 20mm cannon (the museum's plate labels it as for a Type 97, but this is an error), the 20mm shell for the Ho-5, two different shells for the Navy Type 3 12.7mm gun, and the Ho-103 12.7mm round at the bottom.

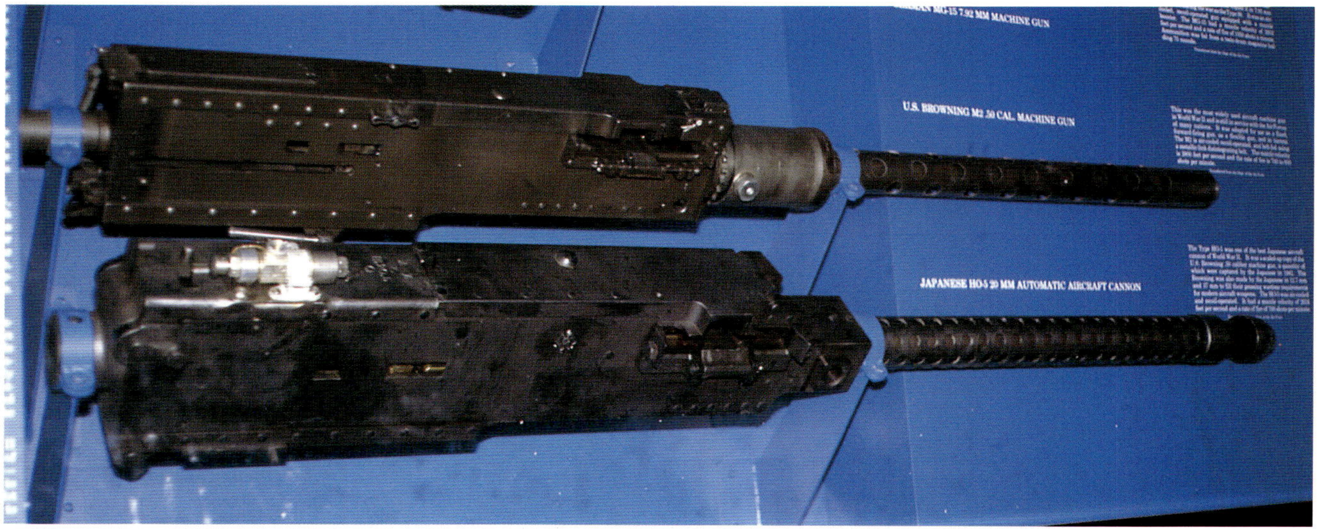

86

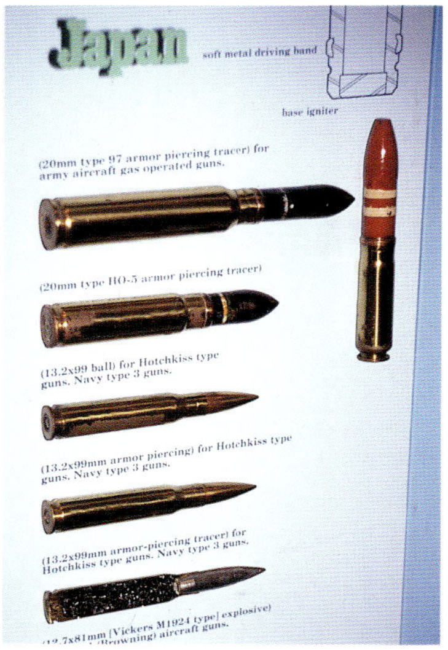

87

88

89

90

91

写真88-89：主翼のホ五機関砲点検用ハッチをみる。写真88は右翼、写真89は左翼。その横にある小判型のちいさなハッチが主翼燃料注入口の外蓋。

写真90：各パネルを閉じた状態の左主翼上面。疾風の主翼外板は必要とされる強度の違いなどから各部で外板の厚みが異なり、0.5、0.6、0.8、1.0、1.2㎜のものが使用された。

写真91：左主翼下面のホ五 20㎜機関砲空薬莢と保弾子の排出口。排出口には主脚と連動するシャッターがある。これは脚下げ時に閉じられて砂塵が砲室に侵入するのを防止する仕組みだが、写真のシャッターは完全に閉じきっていない。排出口の横に増加タンク兼爆弾懸吊ラックがみえるが、その前後に装備されていたゆれ止めは失われている。なお、展示機の爆弾は参考にならない。

Photos 88-89: The access hatch for the Ho-5 cannon on the wings. The small hatch just inboard of that is the wing fuel fill cover.

Photo 90: The upper surface of the left wing with the various hatches seen closed.

Photo 91: The Ho-5 20mm cannon shell and clip ejection ports on the bottom of the left wing. The Hayate had a dust cover mechanism connected to the landing gear which would cover these ports when the gear were lowered to prevent dust and dirt from entering the wing through them. However, in the photo the ports are not completely closed. The bomb and drop tank rack is installed, but the sway stoppers have been lost. Note that the bomb in this photo is completely inaccurate.

主翼　WINGS

92

93

写真92：ファウラー式蝶型フラップと補助翼。中島独得のフラップは全開状態。上面のガイドレール溝がわかる。フラップ後縁のディテールにも注目。この段になった部分はフラップを閉じたときに、主翼後縁ラインからすこしはみ出す（写真113参照）。

写真93：後方からみた蝶型フラップ。蝶型という名称は、作動時にフラップの胴体側が翼端側を支点に、弧を描くように開くため。その動きが羽をひろげる蝶のようすに似ていることから（実際の蝶の羽とは逆方向の動きをするが）つけられたもの。

写真94：補助翼のディテール。操作ロッドとホーンの結合部にはカバーがつく。手前にはヒンジの一部もみえる。

写真95：フラップ全開状態の左主翼下面。フラップ手前のパネルは向かって左が翼内砲点検用、右がフラップ操作装置点検用。

写真96：フラップ操作装置点検パネル裏側のディテール。

94

Photo 92: A view of the "butterfly" combat flaps and aileron on the right wing. The flaps are completely lowered in this view. Note the small fairing for the flap guide rail on the trailing edge. When the flaps are raised, a small portion of their trailing edge protrudes from the wing (see photo 113).

Photo 93: The flaps seen from behind the plane. The name "butterfly" was applied because of the way the inner edge of the flaps drop down using the outboard edge as a pivot, similar to the wings of a butterfly as they open (although a butterfly is in the reverse configuration, with the outboard edge dropping farther than the inboard).

Photo 94: A close-up of the aileron hinge and control rod connection.

Photo 95: The underside of the left wing with the flaps seen fully lowered. The two access panels seen in front of the flap are a wing armament access (left as seen in the photo) and the flap control mechanism access (right).

Photo 96: A close-up of the flap control mechanism access hatch's interior side.

95

96

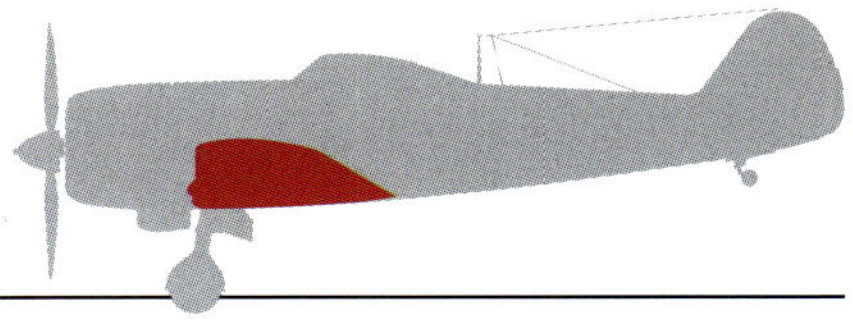

97

98

99

写真97-100：蝶型フラップのディテール。原理的にはファウラーフラップだが、離着陸以外に戦闘中も「空戦フラップ」として使用できる仕組み。疾風以前に中島の一式戦隼、二式単戦鍾馗が、旋回性能向上のために採用したものを踏襲している。写真98はフラップ収容部とガイドレールのディテール。

写真101：真横からみた補助翼とフラップ。フリーズ式補助翼前縁の断面形に注目。補助翼上面より下面の幅（弦長：コード）があるようにみえるのは、断面がこのような形状をしているため。回転軸より前縁部分にはマスバランスが内蔵され、補助翼の重心を回転軸中心より前に置くことでフラッター現象を防止し、同時に操舵力を軽減する。

Photos 97-100: Close-ups of the flaps. Though used for normal take-offs and landings, they were also employed as “combat flaps” during air-to-air engagements. These are a very similar design to those employed on the Type 1 Fighter “Hayabusa” and Type 2 “Shoki,” deployed to boost the aircraft’s maneuverability during dogfights. Photo 98 is a good underwing view of the flap storage area and guide rail.

Photo 101: The flaps and aileron seen from directly abeam the aircraft. The photo reveals well the cross-section of the aileron. This shape makes it appear that the bottom of the aileron has a longer chord than the top.

100

101

主翼　WINGS

102

103

写真102-103：左主翼上面をみる。疾風の主翼平面形は前縁が一直線で、後縁に前進角のあるテーパー翼。しかし、翼厚の最大位置は翼端へいくにしたがって前方に移動するため、実質的に前進翼となっている。これは中島が九七式戦闘機から採用してきた平面形であった。写真102で、着陸灯の外側にみえる面積の広いパネルは、翼前縁に装備する着脱式燃料タンクカバーを兼ねている。写真103は左主翼端。翼端灯（赤）のあるべき部分がパッチで塞がれている。

写真104-105：着陸灯と左主翼前縁。疾風の上反角は6度。着陸灯の消費電力は240W、機体の電球は失われている。なお写真の銃身はホ五とは異なる展示用ダミー。なお、実物のホ五 20mm機関砲は、機軸線に対し1°30′上方に向けて取り付けられていた（ホ一〇三も同様）

Photos 102-103: Views of the Hayate's left wing. The Ki-84's wing leading edge was of a simple, straight-line configuration, but the wing's thickest point moves forward the farther outboard of the aircraft you go, making it a reverse-swept wing in terms of performance. This design was employed by Nakajima beginning with the Type 97 fighter. The wide panel visible outboard of the landing light covered the removable leading edge fuel tank. Photo 103 is of the left wingtip. The location where there should be a (red) wingtip navigation light has been patched with a small oval panel.

Photos 104-105: The landing light and leading edge of the left wing. The Hayate's wing had an upward dihedral of 6°. The landing light burned 240 watts. This aircraft has lost the bulb. Note that the cannon barrel is a mock-up.

104

105

106

108

107

109

110

写真106：機首側からみた左主翼前縁。疾風の翼面荷重は約180kg/㎡あり、これはメッサーシュミットBf109G型の翼面荷重（約195kg/㎡）に迫る値だった。ちなみに、以前の型より機体が重くなったことから搭乗員の不評をかった、零戦五二型の翼面荷重は約130kg/㎡にすぎなかった。

写真107：左主翼下面のディテール。着陸灯は下向きに角度をつけて取り付けられているため、ガラスカバーは前縁上方より、下面側のほうが広い。主脚後方のふくらみは、主脚出し入れ装置のアームをクリアするためのもの。そのかたちは意外に複雑である。

写真108-110：左主翼の翼端まわり下面をみる。写真108のワンタッチ式ロック付きハッチはピトー管の配線点検口。その右、翼端側のハッチ内には繋留索を止める繋止金具がある。写真109はピトー管配線点検口のクローズアップ。写真110は翼端の下面。下面の翼端灯も失われている。

Photo 106: The left wing's leading edge, seen from the fuselage side. The Hayate had a wing loading of 180kg per square meter, roughly in the same class as the Bf109G, which was 195kg/m². As a side note, the Zero Model 52 was not well-liked by its pilots compared to earlier models of that fighter because it had become too "heavy." The Model 52's wing loading? Just 130kg/m².

Photo 107: The underside of the left wing, around the landing gear. The bulge just behind the gear strut bay is to provide clearance to the gear control arm.

Photos 108-110: The underside of the left wingtip. The panel secured with a one-touch fastener in photo 108 is to provide access to the pitot tube's wiring. Photo 109 is a close-up of this hatch. Photo 110 shows the underside of the wingtip, which has also lost its navigation light.

111

112

113

114

写真111-112：補助翼のディテールをみる。補助翼はフリーズ式で、金属骨組みに羽布張り構造。地上調整用の固定タブは左右の補助翼にある。

写真113：主翼後縁のフラップ収納部まわりをみる。フラップ後縁の収容時に主翼から少しはみ出す部分は、大戦中赤で塗られていた。

写真114：主翼下面の後縁部、補助翼とフラップ下面のディテールをみる。補助翼の下面にならぶハトメ状の穴は、Bf109などにもみられる方式の水抜き用。

Photos 111-112: Close-ups of the ailerons. These were of a fabric on metal construction and had fixed trim tabs on either side.

Photo 113: The trailing edge, around the flap retraction area. Note how the flaps project out slightly from the trailing edge even when retracted.

Photo 114: The underside of the aileron and flaps. The small holes visible on the underside of the aileron are water drains, like those that can also be found on the Bf109 and other aircraft.

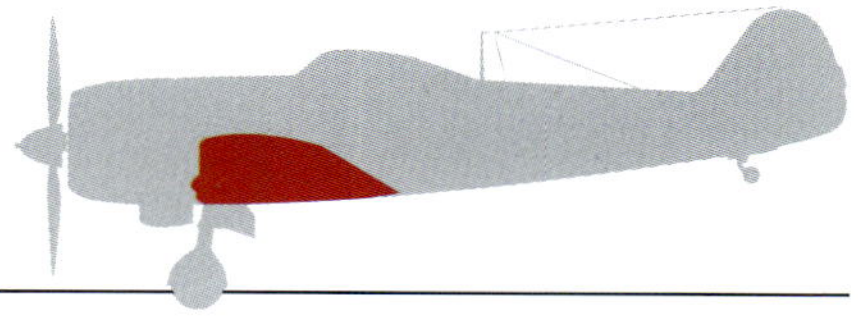

115

写真115：主翼付け根から胴体にかけての構成をみる。前部胴体と一体に製作された主翼付け根は、写真のようにフィレットで覆われる。このフィレットは数枚の板を加工したものを組み合わせ、曲面を構成している。

写真116：後方からみたフィレットの曲面は、後方にいくにしたがってなだらかに変化し、意外と複雑な形状になっている。フィレットのパネルラインにも注目。

写真117：フラップのガイドレール・カバー。写真は左主翼後縁のもの。4本あるフラップのガイドレールのうち、内側の2本にはこのカバーがつく。カバー後端には先のとがった木片のガードがついていたはずだが、この機体では失われている。

Photo 115: A view of the left side of the fuselage, covering the entire wing root. The forward fuselage and wing are of a one-piece construction. Note the multiple-panel construction of the root fillet.

Photo 116: A view of the fillet from behind the wing. Note the surprisingly complex design of the fillet, the curve of which increases in diameter towards the rear of the aircraft.

Photo 117: A close-up of the flap guide rail cover on the trailing edge of the left wing. Two of the four guide rails on the aircraft have these covers at the trailing edge. A pointed, wooden guard should be visible at the end of the rail, but it has been lost on this aircraft.

116

117

主翼 WINGS

118

119

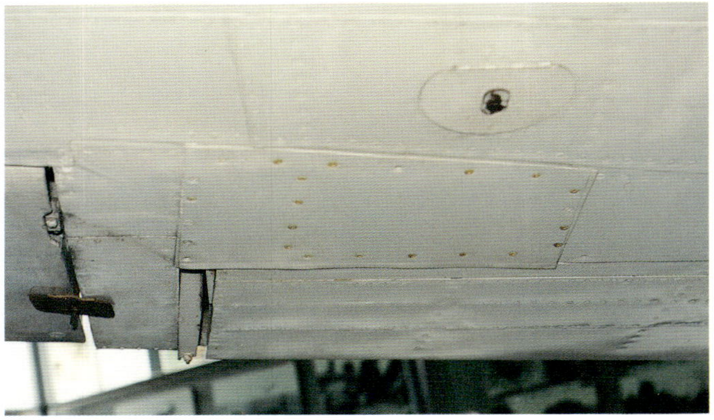
120

121

写真118：右主翼上面をみる。疾風の主翼は後方に補助桁を配した単桁構造。これに対し、隼、鍾馗といった中島戦闘機の主翼は、前後に2本の桁を配した構造だった。

写真119：右補助翼上面のディテール。補助翼を固定している金具は展示用のもの。

写真120：右主翼のフラップ下面。写真右上にみえる丸いハッチは、補助翼操作用第2ベルクランクの点検用。

写真121：右主翼前縁を機首方向からみる。左主翼で着陸灯にあたるスペースは、写真銃の収容スペースとなっている。この写真銃は正式には「射撃鑑査写真装置」と呼ばれるもので、主に訓練用の特別装備であった。写真57に、この写真銃のフィルムカウンターと思われる日本製の計器が写っている。疾風の翼圧比は中央が16.5%、翼端で8%あり、高速機としてはやや厚め。

Photo 118: The upper side of the right wing. The Hayate's wings had a single main spar, with one auxiliary spar to the rear. The Hayabusa and Shoki, by contrast, had another auxiliary spar forward of the main one.

Photo 119: A view of the top side of the right aileron. The lock plate visible on the inboard side is, of course, for display purposes only.

Photo 120: The underside of the right flaps. The round hatch visible in the upper right of the photo is to access the no.2 bell crank of the aileron control system.

Photo 121: The leading edge of the right wing seen from the fuselage side. A gun camera could be mounted on the right wing in the area that corresponds to the landing light on the left wing.

122

123

124

写真122：右主翼下面、翼端付近のディテール。翼端側にみえる半開きになったハッチは繋留索止め金具用。

写真123-125：右主翼下面に装備された燃料冷却器。本体は楕円形断面で滑油冷却器同様、蜂の巣状のコアがある。燃料冷却器は、燃料温度の上昇によって生じた気泡がパイプ内で燃料を詰まらせる、「ベーパーロック」現象を防止するための装備。

写真126：翼端方向から右主翼下面をみる。収容状態のフラップ下面全景ディテールと、燃料冷却器の取り付け位置に注目。

Photo 122: A close-up of the underside of the right wing, near the wingtip. The small, partially-opened hatch visible near the wing tip covers the tie-down ring.

Photos 123-125: Views of the fuel cooler installed on the underside of the right wing. Like the oil cooler, it is an elliptical design with a honeycomb core. The device prevented fuel temperatures from reaching points that would cause vapor lock in the fuel lines.

Photo 126: The underside of the right wing seen from the wingtip.

125

126

主脚 MAIN GEAR & GEAR BAY

127

128

写真127-128：右主脚のディテール。疾風の主脚は、エンジンで駆動する高圧ポンプによって出し入れする油圧引き込み式。写真の脚柱は伸縮部に展示用の筒をかぶせてある。また、タイヤはオリジナルと異なる。主脚の長さは重量軽減などの理由から、2000馬力級戦闘機にもかかわらず1000馬力級の隼や零戦とほぼ同じくらいにおさえられている。これが疾風のプロペラ直径を大きくできなかった原因のひとつとなった。

Photos 127-128: Close-ups of the right main gear. The Hayate's gear were raised and lowered by an engine-driven hydraulic system. Note that for display purposes, this plane has a support pipe covering the normally exposed section of the oleo. The tires are also not original equipment. The Hayate's relatively short legs prevented installation of a large propeller, ultimately about the same size as those of its predecessors the Hayabusa and Zero, despite the aircraft's 2,000hp engine.

129

写真129：後方からみた主車輪。写真は左主脚のもの。

写真130：右主脚のカバーを側面からみる。この主脚カバーは脚柱上部・脚柱下部・タイヤ部の三分割。疾風の主脚は重心位置の関係から車輪を主翼の前方に配置する必要があったため、かなり前傾して取り付けられている。

Photo 129: A view of the left main gear from the rear.

Photo 130: A good view of the outer side of the main gear cover. The Hayate's gear cover was divided into three sections, two on the strut and one on the wheel. Note how far forward in the wing the gear are located. This was related to the aircraft's center of gravity, and resulted in the struts angling rather sharply down from the wing.

130

131

132

133

134

写真131：主脚柱頂部の注油気栓（オイルとエアを注入するためのバルブ栓）。

写真132：左主脚正面。伸縮部がオイルで汚れている。主翼前縁の主脚点検カバーがはずされているため、脚柱付け根のディテールをみることができる。

写真133：左主脚内側のディテール。タイヤ部のカバーは脚柱下部カバーとヒンジで結合され、ブレーキ調整時はこのヒンジ部分から下を取りはずす。

写真134：左主脚のカバー外側をみる。

Photo 131: A close-up of the top end of the strut showing the air and oil valve used to replenish strut pressure.

Photo 132: The left main gear seen head-on. The oleo is stained with oil in this photo taken when the aircraft was still flyable. The leading edge inspection panel has been removed, revealing the strut mount within.

Photo 133: The inboard side of the left main gear. Note how the wheel cover and lower strut cover are coupled with a hinge. The wheel cover could be removed for brake maintenance.

Photo 134: A close-up of the left side main gear cover.

主脚 MAIN GEAR & GEAR BAY

135

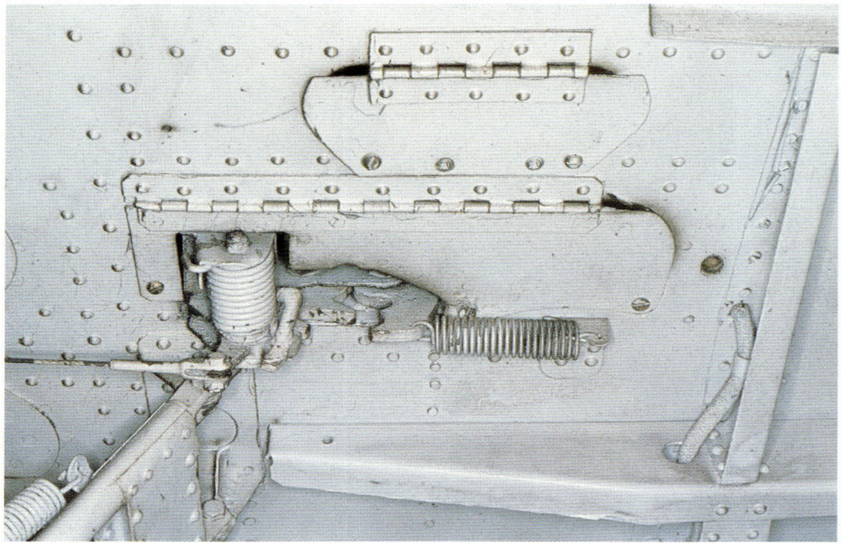

136

137

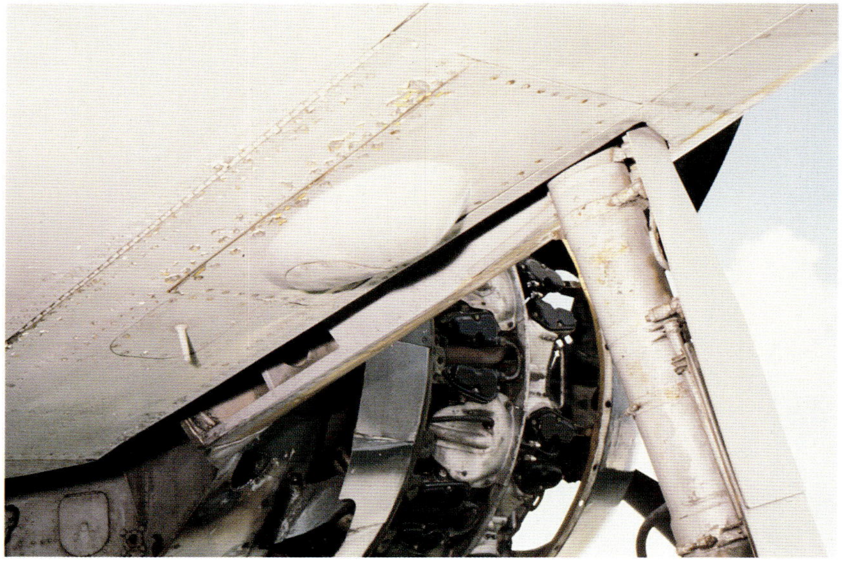

138

写真135：左主脚収容部を見上げる。車輪収容部中央に胴体側タイヤカバーのアップロック装置がある。収容部前方のヒンジ付きハッチはエンジン点検口。

写真136：アップロック装置のクローズアップ。向かって左側にみえるワイヤーでラッチをはずして、ロックを解除する。アップロック装置の上にみえる台形状のヒンジつきハッチは油圧点検口。

写真137：胴体側タイヤカバーを後方からみる。左右カバーのあいだにある開口部は、ホー〇三 12.7㎜機関砲の空薬莢排出口。訓練時にはここに扉をつけて空薬莢を回収できた。

写真138：後方からみた主脚柱の収容部分。主脚出し入れアームのフェアリング、主脚柱収納部にのぞくリブに注目。

Photo 135: A look up into the left main gear bay. Note the locking mechanism visible in the wheel bay. The hinged hatch in the forward part of the bay is for engine system access.

Photo 136: A close-up of the gear locking mechanism. The wire on the left side of the photo releases the lock. The trapezoidal access hatch provides access to part of the hydraulic system.

Photo 137: The fuselage-side wheel covers seen from the rear. The opening between the two covers is the shell ejection port for the nose Ho-103 MGs. During training, a small collection pocket could be attached here to recover the casings.

Photo 138: The right strut bay seen from the right rear.

139

140

141

142

写真139-141：胴体側タイヤカバー内側のクローズアップ。カバーの操作部は「コ」の字型のアームに、スプリング付きの上部操作ロッドを組み合わせた構造。「コ」の字型アームに取り付けられた四角い板を、主脚引き込み時にタイヤが押す。主脚収容口内部の、パイプやワイヤー類などにも注目。

写真142：右主脚の収容部をみる。

Photos 139-141: Close-ups of the inboard tire covers. The covers were closed by a simple mechanism where the retracted tire would push a plate connected to the U-shaped bracket linking the cover to the aircraft.

Photo 142: An overall view of the right side wheel bay.

143

144

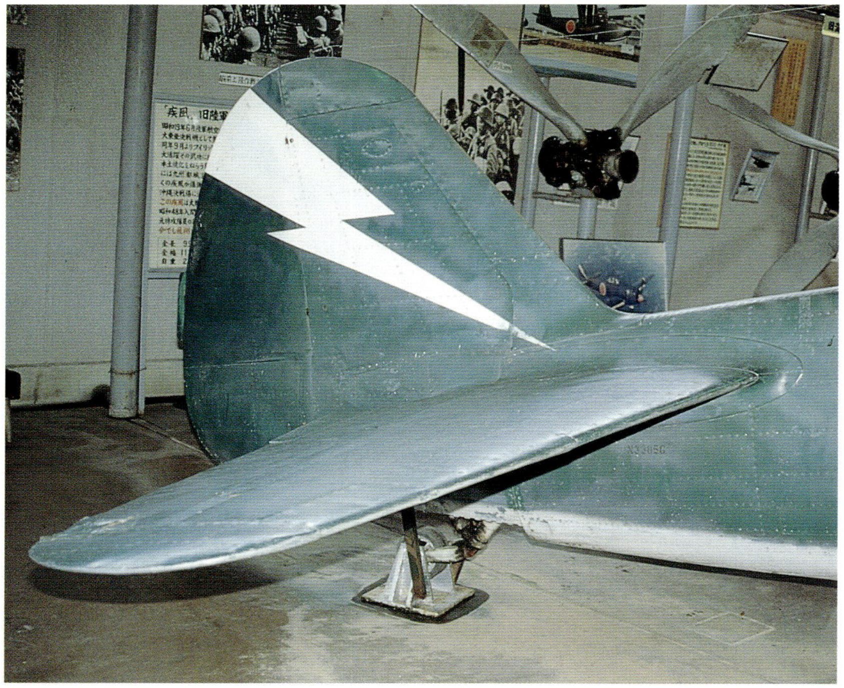

145

写真143：疾風の尾部左側面。尾翼のレイアウトは垂直尾翼を水平尾翼より後方に配置した、中島独得のデザイン。これは方向安定と横安定を別々のものとして考え、垂直尾翼と水平尾翼の距離を離すことで、射撃時の方向安定性をはかったため。二式単戦鍾馗ではじめて採用されたデザインであった。

写真144：垂直尾翼の左側面。垂直安定板は胴体と一体でつくられている。安定板上部にある尾灯は、左右とも失われている。方向舵下部が胴体断面にあわせてふくらむ特徴的なデザインも、中島が隼以来採用してきたもの。疾風は中島が積み重ねてきた技術を集大成した設計の末、誕生した戦闘機といえるだろう。

Photo 143: The left side of the Hayate's tail. The horizontal stabilizer is positioned quite a bit forward of the vertical stabilizer in a typical Nakajima design. This design was first seen in the Type 2 Fighter "Shoki" and improved directional stability when the aircraft used its weapons.

Photo 144: A close-up of the left side of the vertical stabilizer, which was integrated in a one-piece design with the fuselage. The navigation lights that should be present on both sides of the top of the fin have been lost on this plane. The lower part of the rudder bulges somewhat to blend with the fuselage lines.

写真145：右前方からみた疾風の尾部。

写真146：左水平尾翼。尾翼付け根後方にみえる２つの四角いパネルは、尾輪降着装置点検用。水平尾翼前縁外側が前方に突き出している部分には、重量3kgのマスバランスがおさめられている。

Photos 145: The tail of the Ki-84 seen from the right front.

Photos 146: A view of the left horizontal stabilizer. The two small, square panels just aft of the stabilizer root are to access the tail gear mechanism.

146

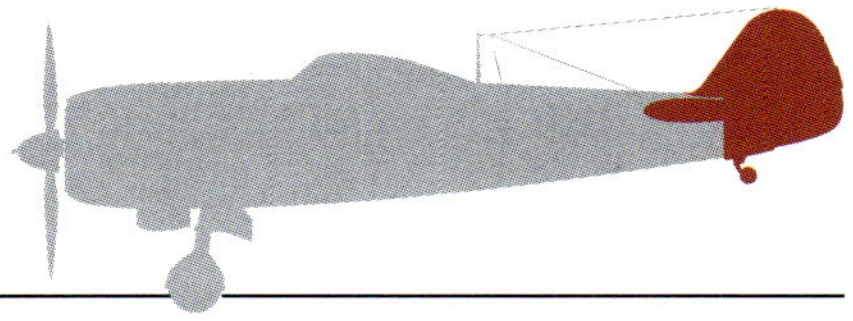

147

148

149

写真147-148：垂直安定板のディテール。安定板の外板は厚さ0.6㎜。尾灯のあった位置は楕円形のパッチで塞がれている。

写真149-150：方向舵のディテール。方向舵は上部が羽布張り、下部が金属外皮構造。後縁には固定タブがつく。尾翼上下全幅にわたる方向舵の構造も、一式戦隼以来のデザインを踏襲している。

Photos 147-148: Close-ups of the vertical stabilizer. The outer panels were of 0.6mm thickness. The navigation light mounts, now lost, have been covered by elliptical panels.

Photos 149-150: Close-ups of the rudder. The upper end of the rudder was fabric-covered, while the lower end was metal-skinned. A fixed tab was mounted on the trailing edge. The rudder runs the entire length of the vertical stabilizer, a design used beginning with the Ki-43 Hayabusa.

150

尾部 TAIL

151

写真151-153：右水平尾翼を後方からみる。水平安定板は左右一体構造。昇降舵は補助翼同様、金属製の骨組みに羽布張り構造。写真153は修正タブのクローズアップ。コクピット内部左側のハンドルで操作する（写真55参照）。

Photos 151-153: The right horizontal stabilizer seen from the rear. The stabilizer is a one-piece construction left to right. The elevator, like the ailerons, is a fabric-on-metal design. Photo 153 is a close-up of the trim tab. This is controlled with a handle on the left side of the cockpit (see photo 55).

152

154

153

写真154：左水平尾翼を後方からみる。展示中に破損したのだろうか、ツギ当てだらけの昇降舵が痛々しい。

写真155：左昇降舵のマスバランス。

Photo 154: The left side of the horizontal stabilizer. The elevator appears to have suffered quite a bit of damage during display.

Photo 155: This view shows off well the mass balance on the left side of the elevator.

155

156

158

157

写真156-157：疾風の尾輪。おそらくオリジナルのものと思われる。尾輪は主脚同様油圧による完全引き込み式。残念ながら尾輪カバーは失われている。

写真158-159：後方からみた尾部。方向舵下面のふくらみ具合がよくわかる。方向舵後縁も金属部分には厚みをもたせてあることに注目。

Photos 156-157: The tail gear of the Hayate. This appears to be original equipment. Like the main gear, it retracted completely via a hydraulic system. Unfortunately, the cover has been lost on this plane.

Photos 158-159: Overall views of the tail of the plane from the rear. These views show off well how the lower part of the rudder bulges slightly to conform to the fuselage lines. Note how the metal frame of the lower portion of the elevator also becomes thicker lower down.

159

四式戦「疾風」の基本塗装＆マーキング
Basic Camouflage & Markings of the NAKAJIMA Ki-84

作図・解説／野原 茂
Illustrations and text by Shigeru Nohara

基本塗装（全面無塗装機）
Basic Markings (Unpainted Aircraft)

左側面 Left side

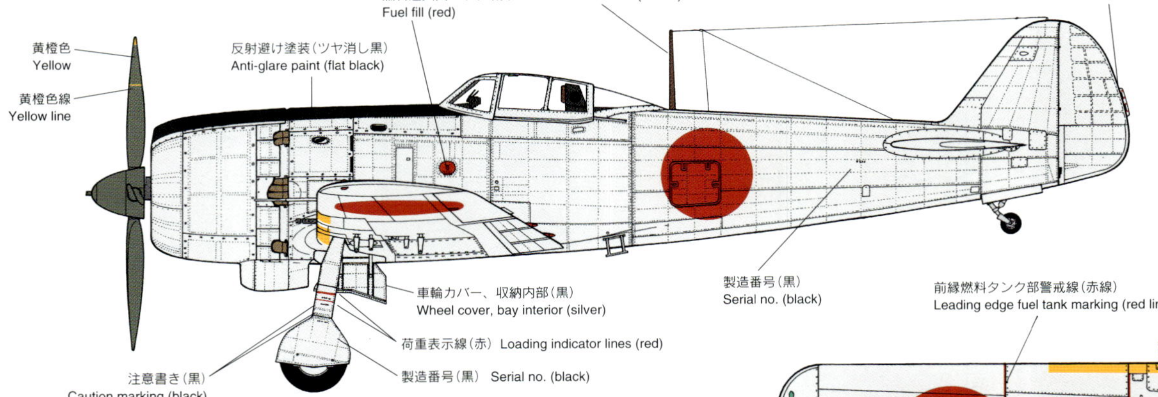

上面 Upper side

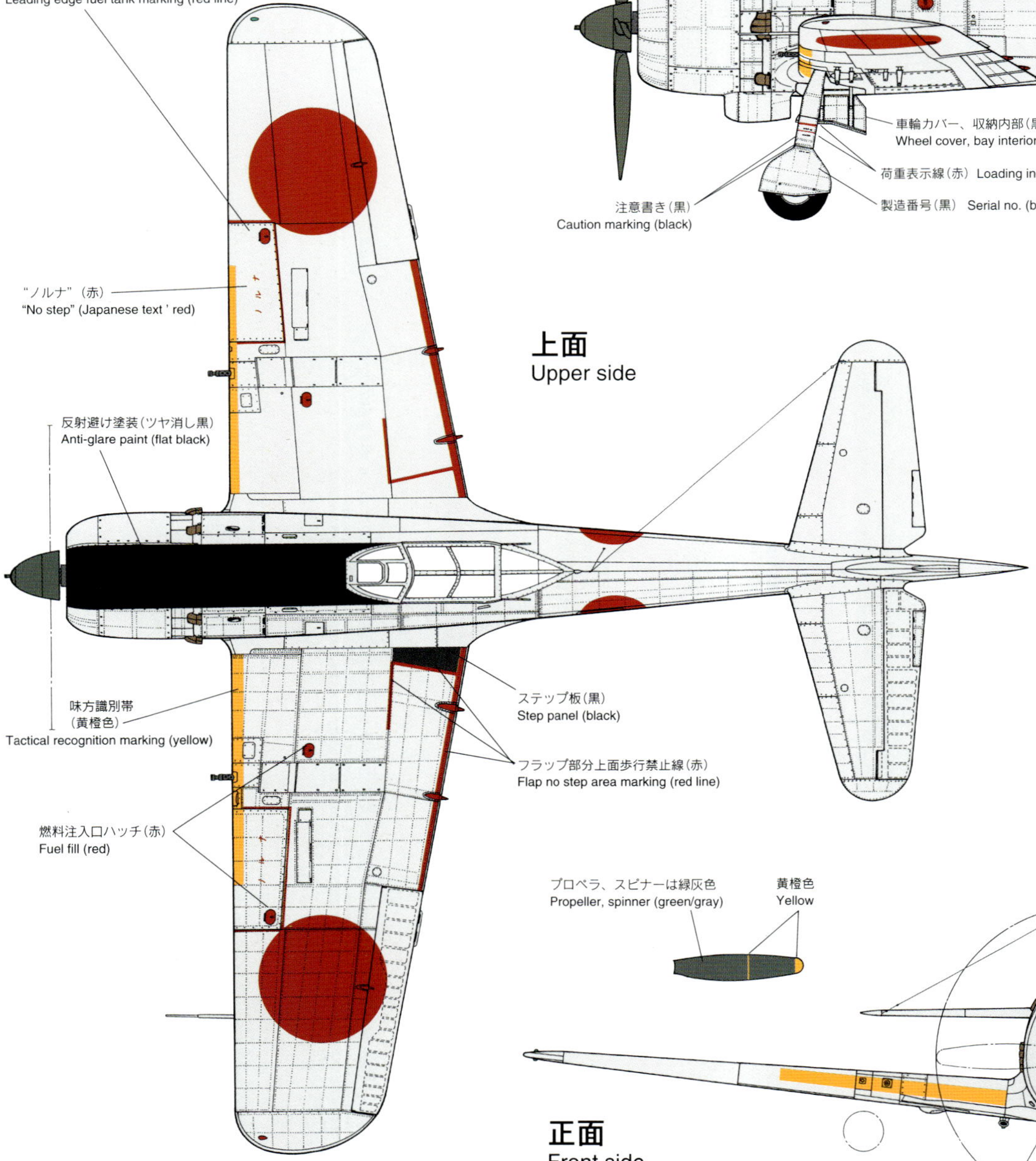

主翼下面 Wing underside

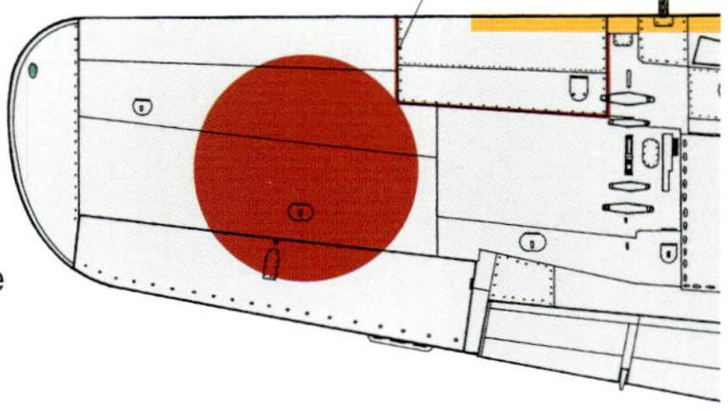

キ84の試作機、増加試作機が完成していた昭和18年中の、陸軍戦闘機の工場完成時点における基本は、全面無塗装（アルクラッド仕上げ）、機首上面に反射除けツヤ消し黒塗装、各日の丸は白フチなしであった。キ84もこれに準じ、昭和19年に入って完成した第二次増加試作機までこれが適用された。補助翼、昇降舵、方向舵の羽布張り部は銀色ドープ（アルミ粉末をワニスで溶いたもの）のようであった。キ43、44、61の各戦闘機は、ハミルトン系プロペラを使っていたこともあり、スピナーも含めてこげ茶色であったが、キ84のみ双方を灰色がかった、少し明るめの緑色、いわゆる緑灰色に塗っていた点が特徴。細部ステンシルなどの要領は図を参照されたい。

During 1943, completed Hayate prototypes and supplementary prototypes were finished unpainted at the time they left the factory. The only significant markings were the flat black anti-glare paint on the upper side of the nose, the yellow recognition stripe on the leading edges of the wings, and the red hinomarus on the fuselage and wings, these without white borders. The second-batch supplementary prototypes completed in 1944 were marked the same way. These aircraft also had aluminum dope painted on the fabric-covered control surfaces (ailerons, rudder, elevator). The Ki-43, 44 and 61, which used a Hamilton-type propeller, had the spinner and prop painted brown, but the Ki-84 alone used a much lighter so-called "green-gray" color. Refer to the illustrations for details on other markings.

正面 Front side

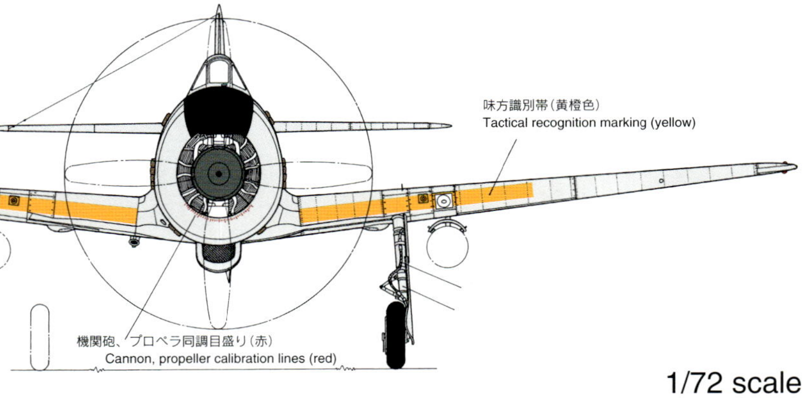

1/72 scale

緑灰色 Green / Gray	赤 Red	黄橙 Yellow	濃緑色 Dark Green	暗褐色 Dark Brown	灰色 Gray

基本塗装(上面濃緑色、または暗褐色／下面灰色迷彩機)

Basic Markings (Upper side dark green or dark brown/underside gray aircraft)

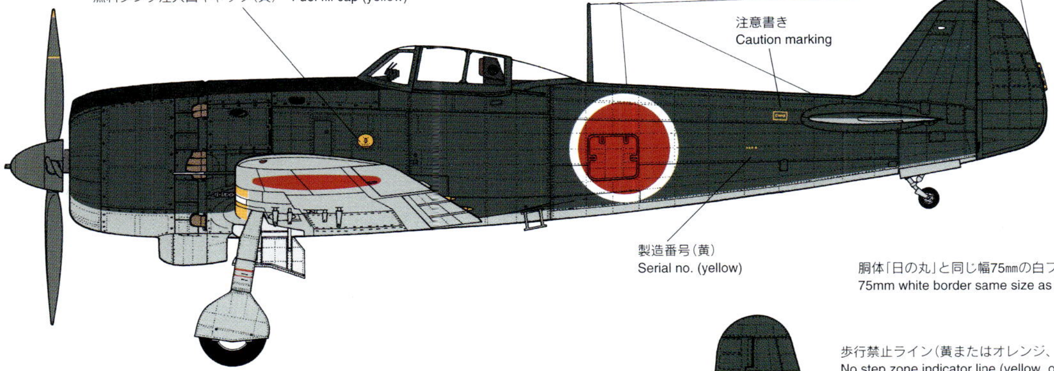

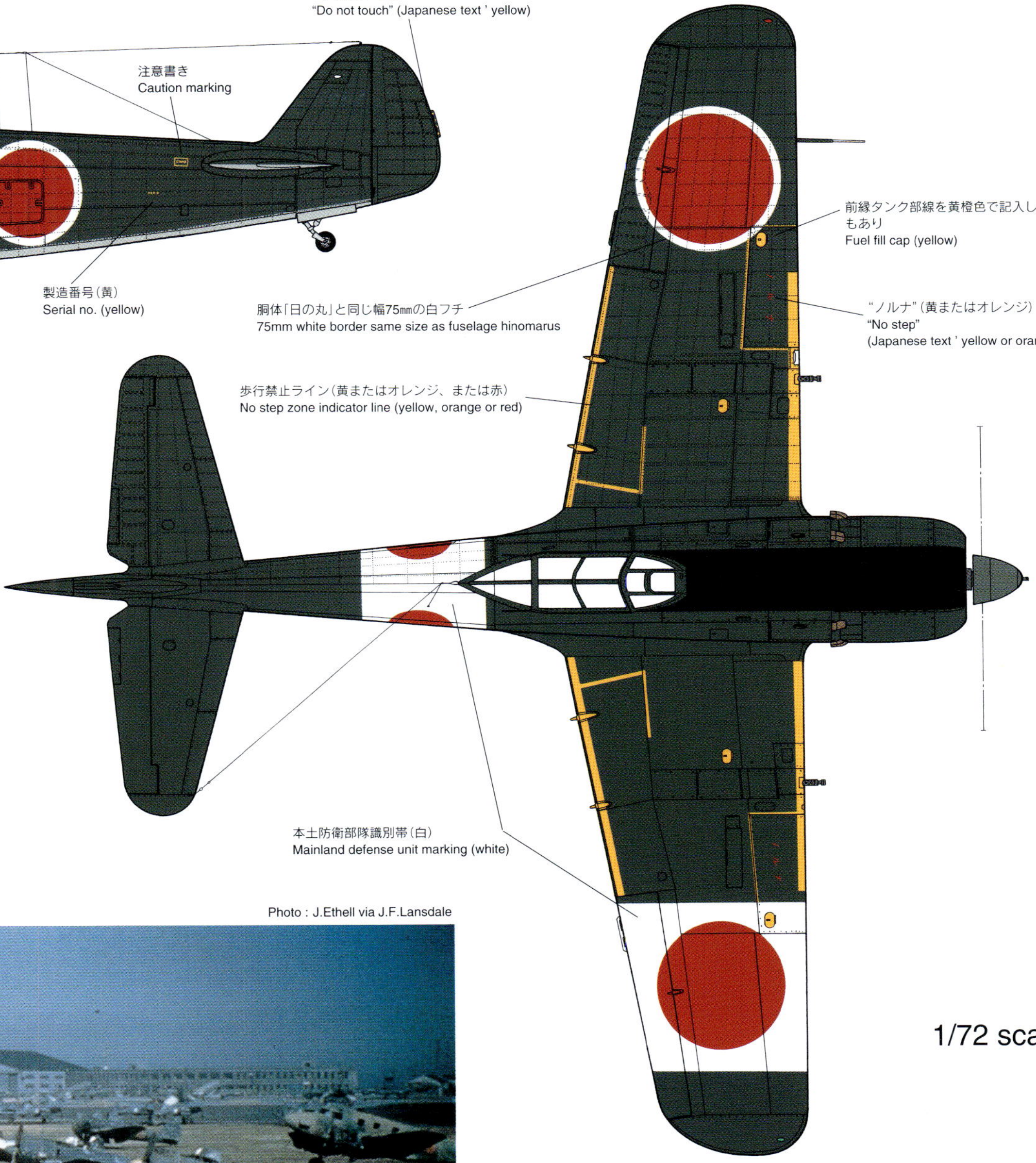

昭和19年3月、第二次増加試作機が最初の実戦部隊飛行第22戦隊に配属されるのと前後し、キ84には迷彩が施されるようになり、当初は無塗装地の上側面に、濃緑色のマダラ、又は蛇行状パターンなどを吹き付けた。しかし、ほどなくして上側面は濃緑色のベタ塗り、下面を灰色に塗るように統一された。迷彩化にともない、胴体、主翼上面日の丸には白フチが追加され、ステンシル、警戒帯なども黄橙色、又は橙黄色に変更された。昭和19年末、陸軍機の迷彩色は暗褐色に統一され、下面もそれにあわせて、やや褐色味を帯びた灰色に変わり、以降敗戦までこれが適用された。

Sometime around March of 1944, when second-batch supplementary prototypes began to be delivered as combat-ready aircraft to the 22nd Sentai, the Hayate began to be finished in a more standard camouflage scheme. Very early examples had the upper surface sprayed in a "snake" pattern or mottled pattern with dark green paint, but this was quickly unified to a solid dark green upper surface with gray underside. Along with this, the hinomarus were changed to include a white border, and caution markings were changed to yellow or orange. Near the end of 1944, IJA aircraft camouflage was unified as dark brown, and the underside gray also changed somewhat towards dark brown. This scheme was used through the surrender.

Photo : J.Ethell via J.F.Lansdale

◀朝鮮の金浦飛行場で敗戦を迎え、焼却処分するために並べられた各種陸軍機。右列はキ43、左列がキ84群で、濃緑色迷彩機と無塗装機が混在している。手前機は、もと飛行第85戦隊第1中隊機、尾翼に"菊水"マークを描いた機は、同22戦隊所属である。迷彩機のなかに、暗褐色塗装機が見当たらないのは、これらの機が19年末以前に生産されたからだろう。いずれにせよ、当時の陸軍機塗装の手がかりとなる、貴重なカラー写真ではある。

Numerous IJA aircraft sit at Kimpo airfield in Korea, waiting to be burned following the surrender. Ki-43 Hayabusas make up the right row, while the left is Ki-84 Hayates, some in green markings, and some unmarked. The plane in the foreground is formerly of the 85th Sentai, 1st Chutai, while the plane with the "Kikusui" marking on the tail was with the 22nd Sentai. The lack of any dark brown aircraft indicates that these planes were all probably manufactured prior to the end of 1944. In any event, this is a rare and valuable photo, revealing the colors of IJA aircraft

キ84甲　飛行第11戦隊第2中隊　昭和20年1月　フィリピン
Ki-84 Ko, 11th Hiko-Sentai, 2nd Chutai, January 1945, Philippines

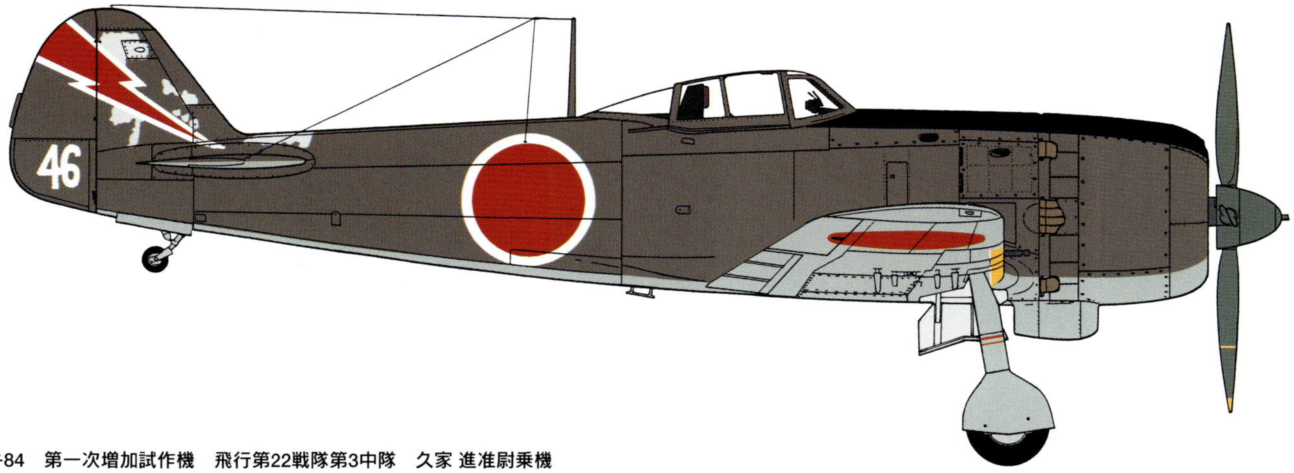

キ84　第一次増加試作機　飛行第22戦隊第3中隊　久家 進准尉乗機
昭和19年4月　神奈川県中津飛行場
Ki-84 First-batch supplementary prototype, 22nd Hiko-Sentai, 3rd Chutai, flown by W.O. Susumu Kuge,
April 1944, Nakatsu AB/Kanagawa, Japan

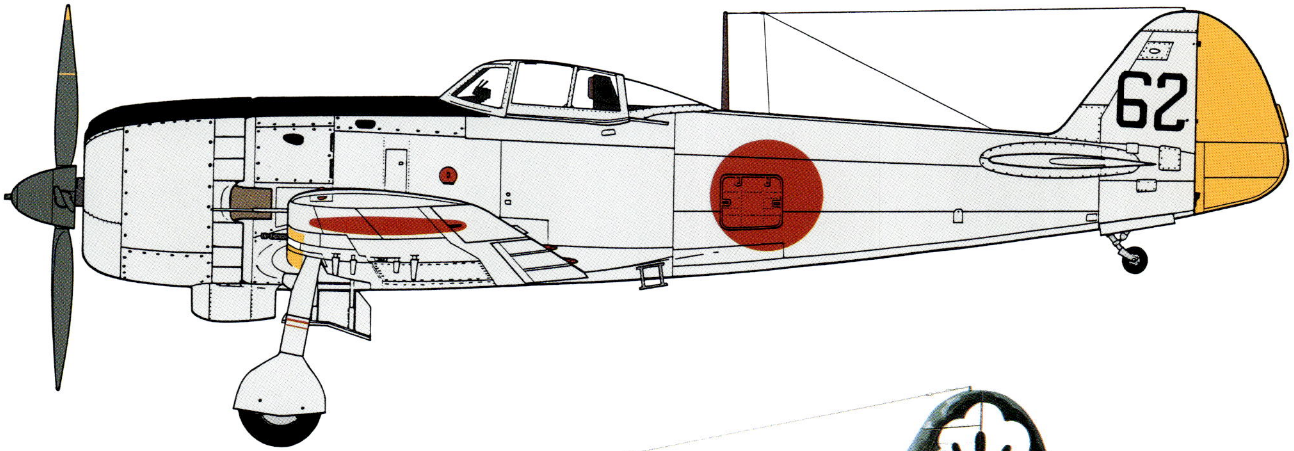

キ84甲　飛行第22戦隊　昭和20年8月　朝鮮／金浦飛行場
Ki-84 Ko, 22nd Hiko-Sentai　August 1945, Kimpo AB/Korea

キ84甲　飛行第25戦隊本部？　昭和20年8月　中国／漢口飛行場
Ki-84 Ko, 25th Hiko-Sentai HQ (?), August 1945, Hankow/AB, China

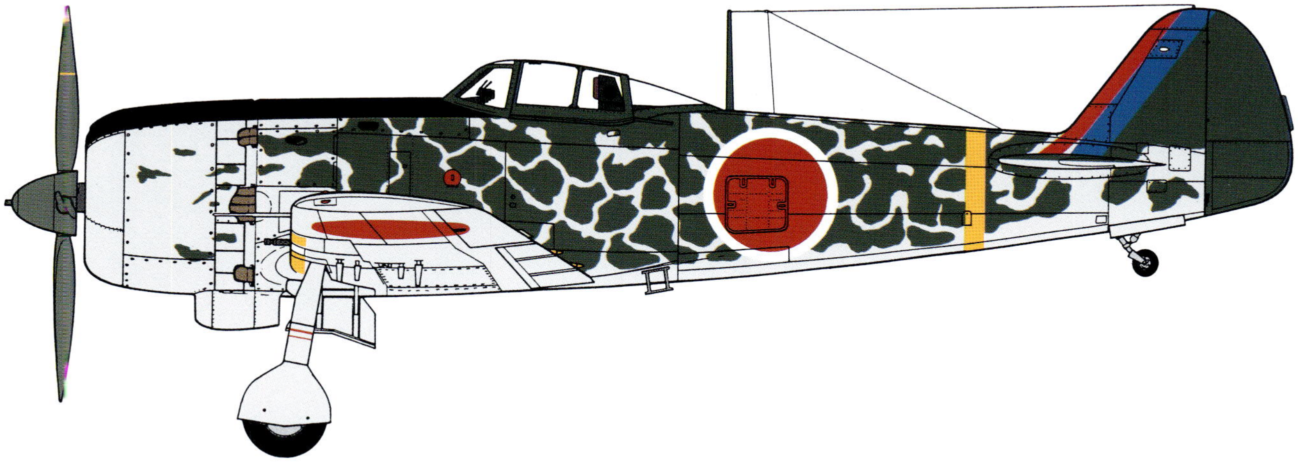

キ84　第二次増加試作機　飛行第47戦隊『桜』隊（旧第3中隊）昭和20年　東京都成増飛行場
Ki-84, Second-batch supplementary prototype, 47th Hiko-sentai "Sakura-tai" (former 3rd Chutai), 1945, Narimasu AB/Tokyo, Japan

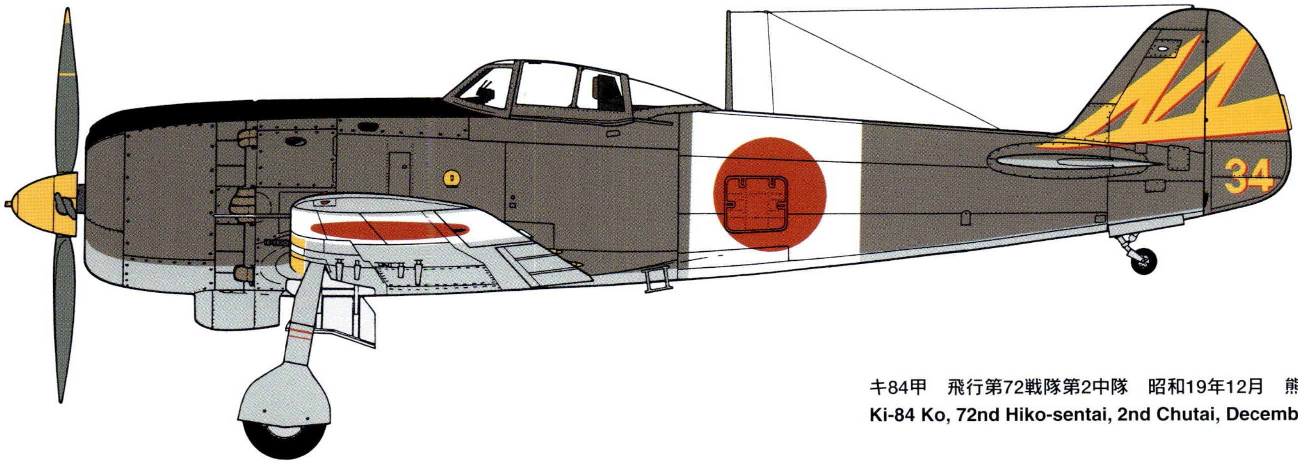

キ84甲　飛行第72戦隊第2中隊　昭和19年12月　熊本県隈ノ庄飛行場
Ki-84 Ko, 72nd Hiko-sentai, 2nd Chutai, December 1944, Kumanosho AB/Kumamoto, Japan

緑灰色
Green / Gray

暗褐色
Dark Brown

濃緑色
Dark Green

灰色
Gray

青竹色
"Aotake" (Interior Color)

黄橙色
Yellow

青
Blue

赤
Red

キ84甲　飛行第101戦隊第2中隊　昭和20年夏　本土
Ki-84 Ko, 101st Hiko-sentai, 2nd Chutai, Summer 1945, Japan

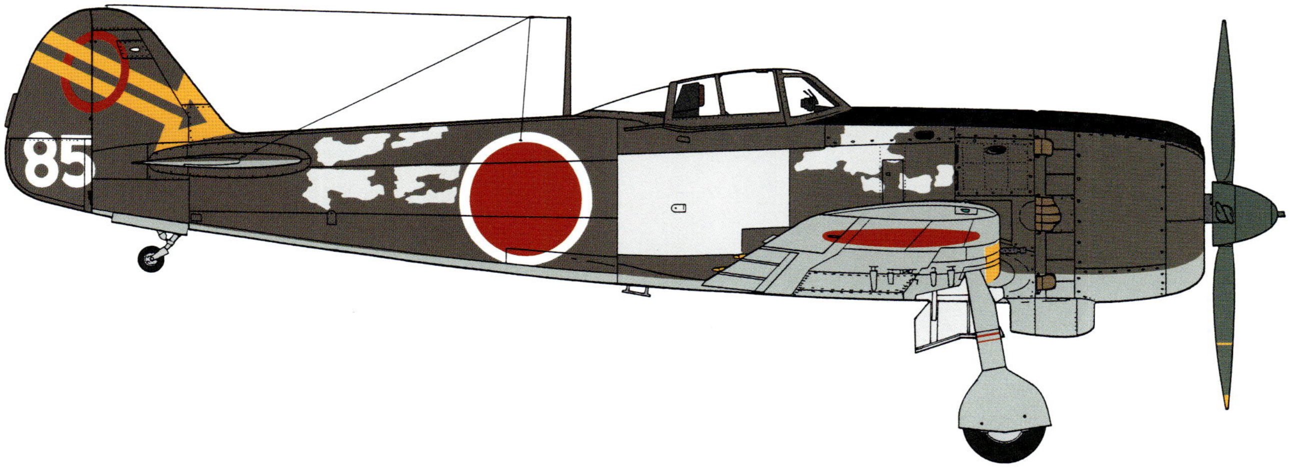

キ84甲　飛行第102戦隊第1中隊　昭和20年4月　宮崎県都城西飛行場
Ki-84 Ko, 102nd Hiko-sentai, 1st Chutai, April 1945, Miyakonojo-nishi AB/Miyazaki, Japan

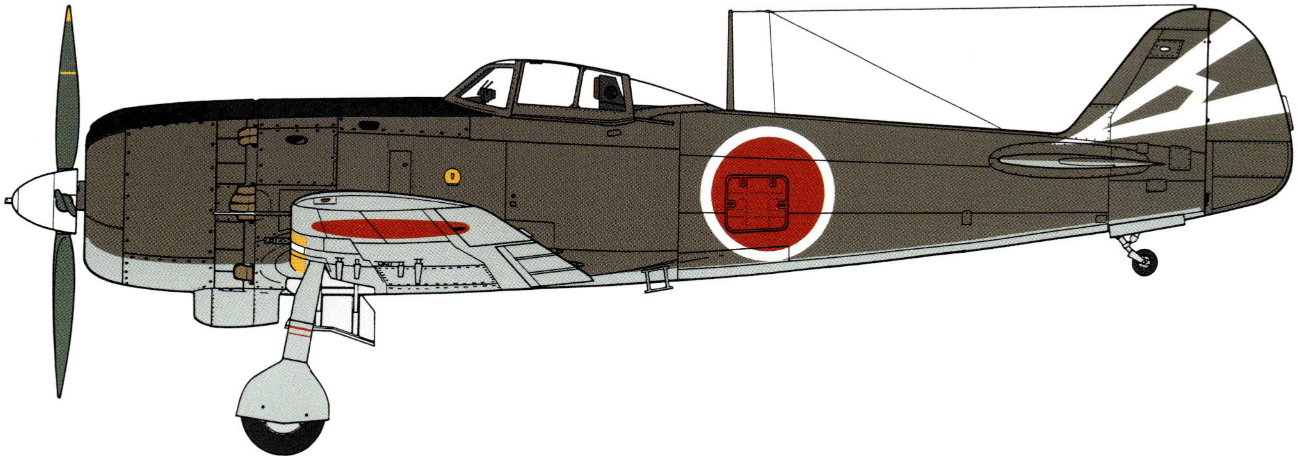

キ84乙　飛行第104戦隊第3中隊　昭和20年8月　旧満州／鞍山飛行場
Ki-84 Otsu, 104th Hiko-sentai, 3rd Chutai, August 1945, Anshan AB/Manchuria

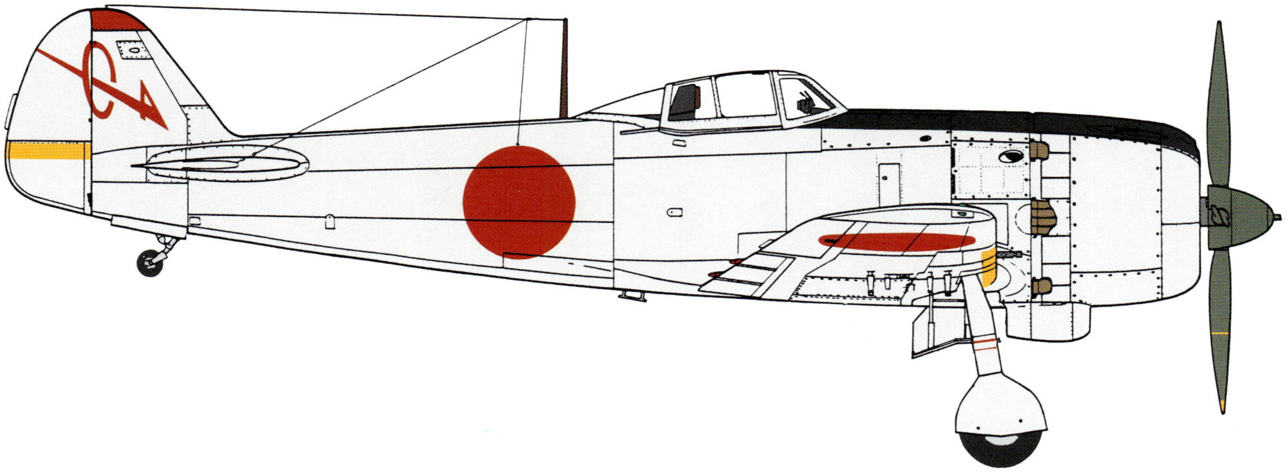

中島 四式戦闘機『疾風』[キ84]

解説・写真提供／野原 茂

Commentary & photo provided by Shigeru Nohara

人の一生に運、不運があるように、軍用機の生涯もまた、それ自体のもつ能力だけではどうにもならない、他の諸々のファクターによって、明暗を分けることがままある。とりわけ、第二次世界大戦のごとき、歴史上未曾有の激戦下に身をさらされた機体は、個々の性能の優劣はともかく、出現のタイミング、運用上の適否などがその実績に大きく影響した。

“大東亜決戦号”という勇ましいキャッチフレーズを冠せられ、全陸軍航空の熱い期待を一身に背負って登場した四式戦闘機『疾風』は、“戦闘機王国”と自他共に認めた中島飛行機（株）が、その全能力を注ぎ込んで送り出した、渾身の一作であった。エンジンは、日本最初の実用2,000馬力級『ハ45』、機体設計は一式戦、二式単戦のそれぞれの長所を採り入れたソツない仕上がりとくれば、陸軍ならずとも、絶大な期待をよせるのも無理からぬことであった。

しかし、残念ながら四式戦は、その期待に応えることができなかった。なぜか？ それは、肝心のエンジンが満足に稼働せず、動いたとしてもカタログデータどおりのパワーが出ず、せっかく機体設計をうまくまとめたことが、生かされなかったことが最大の原因だった。これは、四式戦に限らず、太平洋戦争末期に就役した、同じ『ハ45』（海軍名称は『誉』）エンジンを搭載した海軍機にもいえることで、問題を突き詰めてゆくと、日本陸海軍航空行政全体の思想、国力の限界ということになり、事は簡単ではない。平時ならいざ知らず、戦争という非常時下の兵器の優劣は、単純に設計の良し悪しで論じられるようなものではない。

その好例が、皮肉にも四式戦の心臓たるべき『ハ45』エンジンだった。『ハ45』は、中島飛行機（株）の発動機部が昭和15年（1940）春に設計着手した、わが国最初の2,000馬力級空冷エンジンである。基本的には、一式戦、零戦が搭載した『ハ25』（海軍名称は『栄』）を18気筒化したものだ。構想は的を得ていたのだが、問題は、その出力に見合うサイズの大型化と重量増加を必要以上に抑制し、各部に無理を生じさせたことだ。

緻密、かつ繊細なシリンダー冷却フィン、高回転に耐えるクランクシャフト、マスターロッドなどの製作には高品質の原材料と熟練工の技術を、整備は万全を、使用燃料には高オクタン価ガソリンが不可欠という、いわば、当時の日本では無いものねだりに等しい生産、運用条件を前提にする、あまりに理想ばかりが先行したエンジンだった。試作品のテストが始まった昭和16年3月ごろといえば、もう具体的な太平洋戦争の準備に入っていた時期だ。戦争が始まれば、なおさら上記したような諸条件を満たすのが困難になるのは目に見えている。なんとなれば、太平洋戦争そのものの動機が、南方の石油資源の確保にあったのだから……。陸海軍航空行政責任者は、こんな初歩的、かつ重大なことを見落としていた。『ハ45』は、その計画出力はともかく、戦時下の厳しさに耐えるすべは最初からもっていなかったのだ。

四式戦の部隊就役が進むにつれ、『ハ45』の不調、トラブルは深刻さを増し、例外を除いてほとんどの部隊が40％程度の稼働率にあえいだ。燃料は87オクタンを維持するのが精一杯、性能はカタログデータを下回り、パイロット、整備員の質は低下する一方。数の面でもアメリカ軍に圧倒されるという状況では、もはや本機の活躍など望むべくもなかった。

こんな『ハ45』の不調に、さらに追い打ちをかけたのが、パワー発揮の源でもあるプロペラ。ご承知のように当時の日本では、金属可変ピッチ・プロペラを独自に開発する技術が育たず、太平洋戦争を通じてアメリカのハミルトンスタンダード、およびドイツのVDM製品のコピー、国産化製品、もしくはそれらの改良品でしのいだ。そんななかで、唯一、フランスのラチエ社製の国産化品（ぺ三二）を採用したのが四式戦だった。理由は2,000馬力エンジンに見合うものということだったが、結果的にこの選択は失敗で、本機の性能と実用性を、いちじるしく損ねただけだった。

ラチエは、ピッチ変更を電気式ガバナーによって行う方式だったが、このガバナーの製造がうまくゆかず、ピッチ変更のスムース性を欠き、振動を発生したりした。さらに、主脚を短くして軽量化するという、いわば二義的な理由により、プロペラブレードの直径を異例に小さく抑える（3.05m）という、設計上の大ミスが響き、本来の飛行性能、とくに速度と上昇性能が、そのパワーに見合わない不満足なものになった。アメリカのF6F、F4U、P-47など、同じ2,000馬力級エンジン付単発戦闘機が、直径4m前後のプロペラ（実用性は100％）を採用していたことを考えれば、その差は歴然としている。

いずれにせよ、こうした諸々の悪条件が重なっては、四式戦に華々しい戦績を望むのは酷だろう。ただ、本機の生涯すべてが暗い話題だったわけではない。

四式戦を最初に装備する部隊となった22戦隊は、ベテラン整備員を擁して高い稼働率を保ち、パイロットも粒ぞろいだったから、中国大陸を転戦した1カ月間に、P-51マスタングを含めた在中国米陸軍航空隊、中国空軍を相手に奮戦してかなりの戦果をあげた。また、万全の整備態勢をしいて、つねに100％ちかい稼働率を維持したまれな部隊である47戦隊は、昭和20年に入ってからの本土防空戦で、ときとして目覚ましい戦果をあげている。

『ハ45-21』エンジンの状態がよく、搭乗するパイロットの腕もよければ、四式戦はもっと働けたことはまちがいない。

戦後、米陸軍航空隊のテストで、140オクタンの良質燃料を使い、689km/hの高速を出し、"第二次大戦における日本戦闘機のベスト・ワン"と評価されたことが、本機の開発に関わったひとたちへの、せめてもの救いかもしれぬ。

◀昭和19年11月、フィリピンへの進出を目前に控え、埼玉県の所沢飛行場で報道関係者を対象に、一般公開された飛行第73戦隊第2中隊所属の四式戦『疾風（はやて）』。この日に撮影された写真は、後日に各新聞、グラフ誌などに掲載され、国民は“大東亜決戦号”の勇ましいキャッチフレーズと、『疾風』という愛称を初めて知ったのだった。しかし、皮肉なことに、このころの最前線における実績はサッパリで、その存在感は日に日に色褪せつつあった。

In November of 1944, just before the invasion of the Philippines, the IJA showed this Hayate of the 73rd Sentai, 2nd Chutai at Tokorozawa in Saitama prefecture off to the press and in the following days photos of the plane appeared in nearly every newspaper and photo magazine in the country. Called the “Great East Asian Fighter” this was the first public introduction of the plane with its nickname of *Hayate* (loosely translating as “gale” or other powerful, sudden wind). Despite this aggressive Army propaganda, at the time, the plane was accomplishing very little in actual combat and its presence was hardly felt in some fronts.

四式戦闘機『疾風』実機写真集 “HAYATE” Photo Album

写真提供／野原 茂、押尾一彦ほか Photos provided by Shigeru Nohara, Kazuhiko Oshio

◀昭和17年(1942)4月の設計着手以来、わずか11ヵ月間という短期間の作業で、18年3月に完成へとこぎつけたキ84の試作1号機。のちの生産型にくらべて、それほど大きな違いはないが、主翼だけ面積が19㎡(生産型は21㎡)と小さいのが根本的に異なった。もっとも、このアングルからでは、その違いはわからない。写真は、陸軍における各種審査を終了したのち、中島飛行機に貸与され、落下増槽関係のテスト機として使われていた、昭和19年1月の撮影。

The first prototype of the Ki-84, which went from initial design, beginning in April of 1942, to completion just 11 months later in March of 1943. Differing only slightly from the production models to follow it, the prototype nevertheless had a wing of only 19㎡ as compared to the 21㎡ of production aircraft, though the difference would be virtually indiscernible in a photo taken from this angle. This photo was taken in January of 1944 while the plane was on loan to Nakajima for external fuel tank mounting tests following the completion of all Army testing.

▶試作機3機に続き、合計125機!! もつくられた増加試作機のうち、第一次生産分のごく初期に完成した製造番号102号機。通常、多くても10機前後が原則だった増加試作機を、このように大量発注したのは、陸軍が一刻も速く本機を実用化するため、改修要求を速やかに実施し、量産体制を固めて、部隊の転換訓練を容易にすることなどの理由からだった。写真は昭和19年12月の撮影で、やはり社内テスト用として陸軍から中島に貸与された後のもの。生産機と異なる方向舵の形状に注目。

Following the construction of three prototypes, a remarkable 125 supplementary prototypes were built in two batches. This photo is of #102, one of the early aircraft from the first batch. Normally, only 10 or so supplementary prototypes would be built, but the IJA wanted to get the Hayate into service as quickly as possible, and the production of a large number of prototypes not only allowed rapid testing and implementation of design modifications, but also relatively quick establishment of production techniques. The photo was taken in December of 1944 while the aircraft was on loan to Nakajima for testing. Note the rudder, which is of a different shape than that of the production aircraft to follow.

解説／野原 茂
Commentary
by Shigeru Nohara

◀東京の福生飛行場に所在した陸軍航空審査部において、実用テスト中のキ84第一次増加試作機の1機。製造番号124号機。カウリング上部の突出した機銃発射口覆、集合排気管、防弾鋼板だけのロールバー、上方弦長の狭い方向舵形状など、第一次試作機のオリジナル形態をよく示している。垂直安定板に記入された製造番号下2桁“24”の書体が、前頁に掲載した“102”号機と同じであることがわかる。

Aircraft #124, from the first batch of supplementary prototypes, undergoing flight testing at the Army Flight Test Center at Fussa in Tokyo. This plane displays all the characteristics of the first batch supplementary prototypes, including the protruding muzzle covers on the top of the cowling, clustered exhaust pipes, a roll bar formed only from armor plate, and the rudder with a relatively narrow upper bulge. Note that the “24” marked on the stabilizer is in the same font as that of “102” on the previous page.

▶キ84第一次増加試作機の機首まわりクローズアップ。そのコンパクトさを、設計上の命題にした『ハ45』エンジンを搭載しただけに、本機の機首まわりは一式戦、零戦など1,000馬力級戦闘機と比較しても、それほどきわだって大きいという印象はない。ただ、日本機には前例がなかった、電気式可変ピッチ機構のラチエ4翅プロペラが、みる者に2,000馬力級の迫力を感じさせる。残念なのは、このプロペラ直径を、異例に小さく押さえるという設計上のミスにより、少なからぬ性能上のロスを招いてしまったこと。

A close-up of the nose of a Ki-84 from the first batch of supplementary prototypes. Thanks to the compact design of the Ha-45 engine, the plane doesn’t appear much larger than its 1,000hp predecessors, such as the *Hayabusa* (Oscar) or the Zero. Nevertheless, the four-blade, electric, variable-pitch propeller was unprecedented for a Japanese aircraft and appears much more powerful than that of other IJA planes. Still, even this propeller was a failure, designed with too small a diameter and depriving the plane of deserved performance.

◀陸軍戦闘機パイロットにとって、いわば登竜門とでもいうべき存在の明野飛行学校に配属された、キ84第一次増加試作機の1機、製造番号84013。カウリング上部の機銃発射口覆、ホイールカバー、主翼中央下面の増槽架、ロールバー、方向舵などは、第一次増試機のオリジナル状態だが、排気管のみ生産機に準じた推力式単排気管に改修されている。全面無塗装ジェラルミン地肌（アルクラッド仕上げ）、垂直安定板の製造番号下2桁"13"は黒、方向舵に明野飛行学校マーク（赤と白）を描いている。

One of the Ki-84 supplementary prototypes, no. 84013, on display at the Akeno Flying School, an elite institution considered the gateway to success for Japanese Army pilots. While the aircraft's muzzle covers, wheel covers, drop tank racks, roll bar and rudder profile are consistent with those of the supplementary prototypes, the exhaust pipes alone appear to have been converted from the cluster type to the single, jet-type of the production aircraft. The aircraft is seen in a natural metal (unpainted alclad) finish with the last two digits of the aircraft no. (13) in black in front of the red and white logo of the Akeno School.

▶愛知県の清洲飛行場に飛来したキ84第二次増加試作機の1機。第二次増加試作機は、主翼中央下面の増槽架を除き、ほぼ生産機に順じた外観となったが、写真の機は、主翼武装、風防正面の防弾ガラスなどが未装着で、実戦部隊配備機ではないようだ。垂直尾翼の"37"（白フチ付き赤）は製造番号（437？）の下2桁。

A supplementary prototype of the second batch seen at Kiyosu airfield in Aichi prefecture. Planes of the second batch, with the exception of the underwing drop tank racks, were nearly identical to production aircraft in appearance. However, this plane is lacking wing armament and bulletproof glass, so it is unlikely that it saw combat unit assignment. The "37" on the tail fin is in red with a white border and represents the last two digits of the aircraft production number (437?).

◀53頁に掲載した写真と同じ日に撮影された、所沢飛行場における飛行第73戦隊のキ84第二次増加試作機。手前の無塗装機は第2中隊、右後方の迷彩機は第3中隊所属である。昭和19年5月に編成された73戦隊は、姉妹隊の71、72戦隊と共に、比較的早い時期の結成部隊ということもあり、生産機に混じって増加試作機も保有していた。ちなみに、写真手前の機体、製造番号"491"号は方向舵上方の弦長が少し大きいものを付けている。

Photographed the same day as the plane in photo on page 53, These are second-batch supplementary prototypes of the 73rd Sentai seen at Tokorozawa. The unpainted aircraft in the foreground is of the 2nd Chutai, while the camouflaged plane in the right background is assigned to the 3rd Chutai. Formed as one of the earliest Hayate air groups along with the 71st and 72nd Sentais, the 73rd had as a result a number of supplementary prototypes mixed in with its roster of production aircraft. Note that the foreground aircraft (no. 491) appears to have a rudder with slightly longer chord than normally seen.

◀昭和19年3月、福生飛行場から離陸する直前の、飛行第22戦隊第1中隊所属、舟橋四郎少尉搭乗のキ84第二次増加試作機、製造番号“932”。22戦隊は、キ84を配備された最初の実戦部隊として知られ、このあと、19年8月から中国大陸に移動して、中国空軍、在中国米陸軍航空隊を相手に奮戦する。尾翼には、有名な“菊水”の部隊マーク（赤いシャドー付きの白）が描かれている。

A view of production number “932,” a second-batch supplementary prototype, seen moments before takeoff from Fussa with 2nd Lt. Shiro Funahashi of the 1st Chutai, 22nd Sentai at the controls, March 1944. The 22nd Sentai was the first combat unit to receive the Hayate and in August of ’44 the group moved to China where they faced Chinese as well as American units. The famous *Kikusui* marking (in white with a red shadow) can be seen on the tail.

▼昭和20年初夏、清洲飛行場の土盛りの掩体内で翼を休める、キ84第二次増加試作機。当時の日本陸海軍機が、米軍機に対する視認率を下げるため、ほとんど日の丸標識の白フチを廃止していたなかで、胴体、主翼上面とも同じ幅の白フチどりで、くっきりと目立たせているのが印象的。清洲飛行場は、当時二式複戦（キ45改）から五式戦（キ100）に機種改変中の、飛行第5戦隊の本拠地だったので、写真のキ84は、名古屋周辺の飛行場から迎撃戦にでも出動し、不時着したものだろう。

A second-batch supplementary Ki-84 rests its wings at Kiyosu in early summer, 1945. At the time, nearly all Japanese aircraft had the white border around the red *hinomaru* rising sun markings removed to improve the effectiveness of their camouflage, but this plane’s circles remain exceptionally prominent. At the time, Kiyosu was in use by the 5th Sentai, which was converting from the Type 2 Two-seat Fighter (Ki-45 Kai) to the Type 5 Fighter (Ki-100). As such, this Ki-84 probably was forced to land here for some reason after an action against U.S. forces that originated at another airfield around Nagoya.

◀調布飛行場を本拠地にした三式戦（キ61）装備の飛行第244戦隊と共に、帝都防空の陸軍戦闘機隊の中核を成した、成増飛行場の飛行第47戦隊所属キ84。本土防空部隊を示す、日の丸周囲の白帯も鮮やかな手前機は、第二次増加試作機の1機。画面左端に、隊名の“47”を図案化した部隊マークを描いた隣の機の尾翼が写っている。色は黄のようなので、『桜』隊（旧第3中隊）と思われる。

Here's a Ki-84 of the 47th Sentai at Narimasu, which together with the Type 2 Fighters (Ki-61) of the 244th Sentai at Chofu formed the core of the IJA's Tokyo Air Defense forces. The white band around the fuselage *hinomaru* signifies a Tokyo Air Defense aircraft, in this case a second-batch supplementary prototype. The vertical stabilizer of another aircraft with the stylized "47" marking of the unit can been seen at the left edge of the photo. The color appears to be yellow, which would make the plane part of the so-called "Sakura Squadron" (Sakura-tai), the former 3rd Chutai.

▶主翼下面に、200ℓ入統一増槽を懸吊し、フィリピンのルソン島クラーク飛行場から出撃する、生産型キ84甲。尾翼マークを未記入のため、所属部隊は明らかではない。昭和19年10月に始まったフィリピン攻防戦は、キ84にとって最初に経験する大きな戦いであり、実に9個戦隊が投入された。しかし、稼働率の低さ、パイロット練度の低さ、作戦上のミスなども重なり、圧倒的に優勢な米軍を相手に、ほとんどの部隊が全滅に近い損失をこうむった。

With 200-liter drop tanks under the wings, a production Ki-84 Ko prepares to takeoff from Clark Field in the Philippines. Bearing no tail markings, the plane's unit affiliation cannot be determined. The air defense of the Philippines, beginning in October of 1944, was the first large-scale, sustained combat action seen by the Ki-84 and no less than nine Chutais were involved. However, persistent maintenance problems, pilot inexperience and strategic errors resulted in most of these units suffering huge losses at the hands of American air forces.

◀奮戦空しく、フィリピンの飛行場（クラーク？）で機体をジャッキ・アップしたまま、進攻してきた米軍に接収された、もと飛行第72戦隊第3中隊のキ84甲“81”号機。尾翼の横帯が72戦隊マークで1中隊：白、2中隊：赤、3中隊：黄で記入した。写真の“81”号はカウリング先端とスピナーも中隊色の黄に塗っている。72戦隊もまたフィリピンにて壊滅的な損害をこうむった部隊で、戦力回復の目途が立たず解散に追い込まれた。

Ki-84 Ko no. "81" of the 3rd Chutai, 72nd Sentai, seen up on jacks at Clark (?) in the Philippines after capture by advancing U.S. forces. The horizontal band on the tail was the marking of the 72nd Sentai, with 1st Chutai aircraft marked in white, those of the 2nd in red, and the 3rd in yellow. "81" has its spinner and forward edge of the cowling marked in its unit's yellow as well. The 72nd Sentai was essentially wiped out in the Philippines and with no way to replenish it, the unit was disbanded.

◀宇都宮にて米軍に接収され、本国ライト・フィールドに運ばれて、入念に調査、テストされたキ84甲、米軍登録コード"T2-302"号機。戦闘重量での比較審査では、高度6,000mでP-51、P-47をしのぐ最高速度を記録して米軍をおどろかせた。しかし、テストは排気管の材質および溶接技術の不良からくるトラブルにより、しばしば中断。排気管の材質と工作は最低と評価された。

One of the two Ki-84 Kos captured by American forces south of Utsunomiya seen in the U.S. after extensive physical and flight-testing. Powered by high-grade 130-octane gasoline, something Japanese engineers could only dream of, the plane is said to have reached 689km/h, a full 60km/h faster than any Japanese tests, surprising its American evaluators with its remarkable performance. The Hayate was considered equal to or superior than the U.S. P-51K and P-47N in early climbing performance, level-flight speed and overall maneuverability. It received negative marks for pilot protection and range and also suffered numerous exhaust pipe problems during testing. This Hayate is believed to have eventually been scrapped in the U.S.

▶群馬県の中島飛行場・太田工場に隣接する太田飛行場に並べられたまま、敗戦を迎えたキ84群。手前から2～4機目は、もと飛行第104戦隊第1中隊所属のキ84乙で、胴体後部に太い白帯を巻いた機は、中隊長乗機だったようだ。キ84乙は機首上部兵装をホ五 20㎜機関砲に換装した、対爆撃機迎撃型だった。104戦隊は、敗戦当時には主力が旧満州(現在の中国東北部)に展開しており、写真の各機は、何らかの理由で中島工場に返納するために戻ってきたらしい。

A cluster of Ki-84s seen outside Nakajima's Ohta Plant in Gunma Prefecture after the surrender. The 2nd to 4th aircraft from the foreground of the photo are Ki-84 Otsu planes, formerly of the 1st Chutai, 104th Sentai. The plane with the large fuselage band was the unit leader's plane. The Ki-84 Otsu was intended as an anti-bomber aircraft and had its nose MGs replaced with the Ho-5 20mm cannon, as in the wings. At the time of the surrender, the 104th Sentai was based in Manchuria, so these aircraft must have been returned to Nakajima for some reason after the end of the war.

四式戦『疾風』ディテール・イラスト＆フォト

イラスト・解説／野原 茂
Illustrations & commentary by Shigeru Nohara

DETAIL ILLUSTRATIONS & PHOTOS

Fuselage rib placement and outer skin thickness
胴体円框（フレーム）および外板厚構成

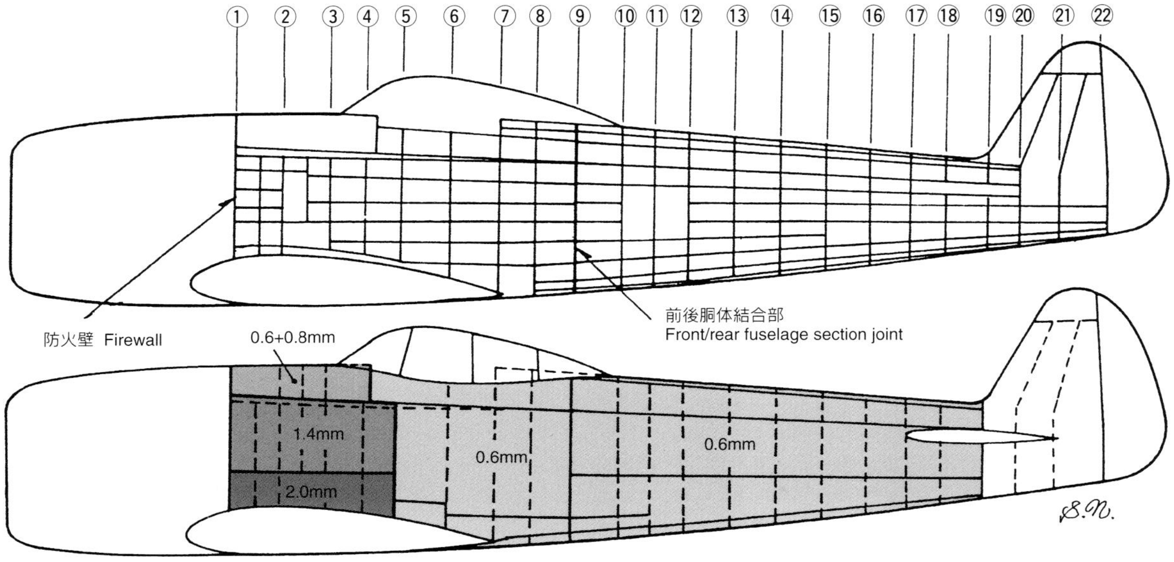

Vertical stabilizer structure (units: mm)
垂直尾翼骨組み図（寸法単位mm）

安定板
Vertical stabilizer

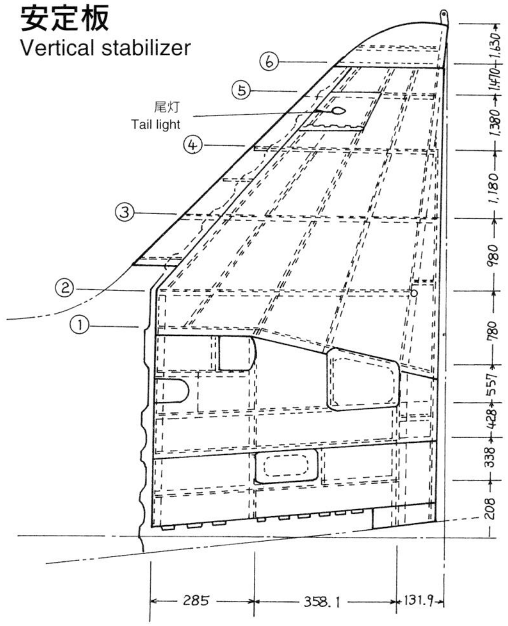

Fuselage structure detail (units: mm. First nine sections shown with outer skin removed)
胴体骨組み詳細図（寸法単位mm） ※第1～9円框間は外板を省いた状態を示す

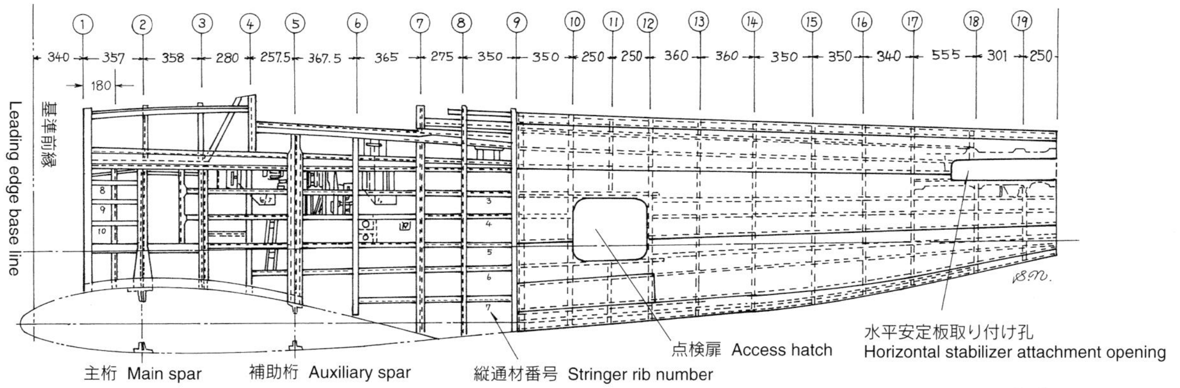

方向舵
Rudder

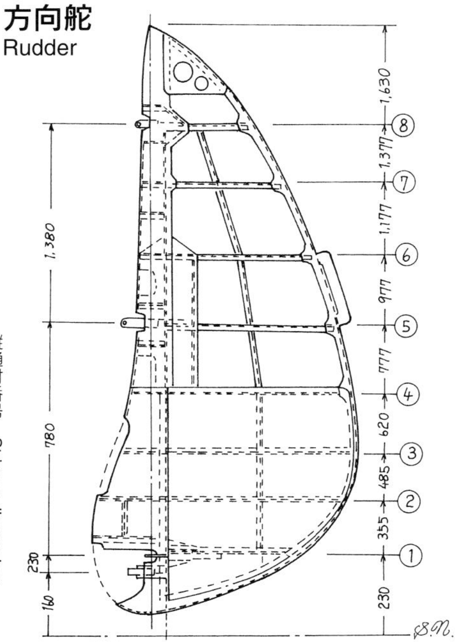

胴体断面
Fuselage cross-section

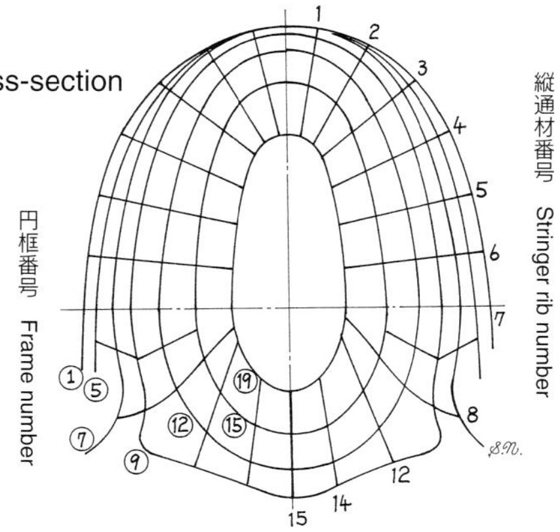

▲キ84の胴体は、基本的にキ27、43、44と続いた中島戦闘機のそれを受け継いだ設計、構造で、とくに目立った特徴はない。22本のフレームに、左右計23本の縦通材を通した骨組みに、区画ごとに0.6～0.2㎜の外板を張った、ごく一般的なセミ・モノコック構造である。第9フレーム部分で前後に分割され、前部が主翼と一体造りになっている点も、キ27以来の中島方式。垂直安定板も後部胴体と一体造りになっており、水平尾翼取り付け位置よりもかなり後方によった配置をとったのは、方向舵操作時のモーメント・アームを長くし、射撃時の機体の"座り"をよくするという、中島独自理論にもとづく処置。

The fuselage of the Ki-84 is of a typical Nakajima design, much like that of the Ki-27, 43 and 44 which came before it. 22 frames are strengthened laterally by 23 stringers in a very typical, semi-monocoque construction. The fuselage was split at the ninth frame into front and rear sections, with the front section being a one-piece construction with the wing, another typical Nakajima design first seen on the Ki-27. The vertical stabilizer is integrated with the fuselage, and the positioning of the horizontal stabilizer well forward of it was a Nakajima design to prevent the two control surfaces from interfering with each other and to stabilize the aircraft better during weapons firing.

Wing structure
主翼骨組み図

▶キ84の主翼もまた、キ27〜キ43のそれに酷似した平面形状で、中島戦闘機の定番といえるものだったが、さすがに内部構造は2,000馬力級戦闘機に見合う内容に進歩していた。主桁、補助桁各1本に24本のリブを配した骨組みで、内翼の主桁〜後桁間に3個、左右外翼の前縁に各1個の燃料タンクを収め、合計697ℓの要領を確保していた。翼厚比は付け根で16.5%、翼端で8%と、このクラスの戦闘機にしては厚いが、要求された航続距離をクリアするための燃料、主脚、20㎜機関砲の収納スペースを確保するため、必然的にこの値となった。

The wings of the Ki-84 were also of a very typical cross-section, along the lines of the Ki-27 to Ki-43 aircraft, but internally, there were important differences related to modifying the structures to be able to take the power of a 2,000hp engine. 24 ribs were installed on a main spar and one rear auxiliary spar. The plane could carry 697 liters of fuel in any one of three tanks located in the inboard wing between the center and rear spars, or in a single outboard leading edge tank. The wing was of a relatively thick design for fighters of its class, but this was necessary to give the plane the internal space needed to store the fuel and armament that specifications demanded.

左翼上面側
Left wing, top view

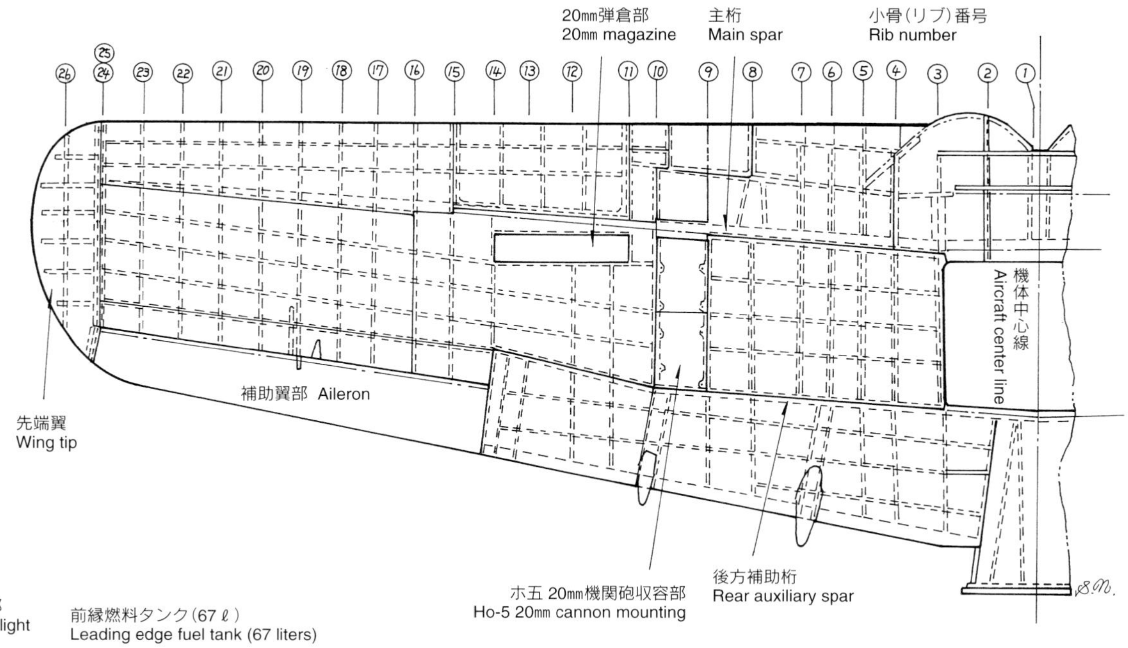

左翼下面側
Left wing, bottom view

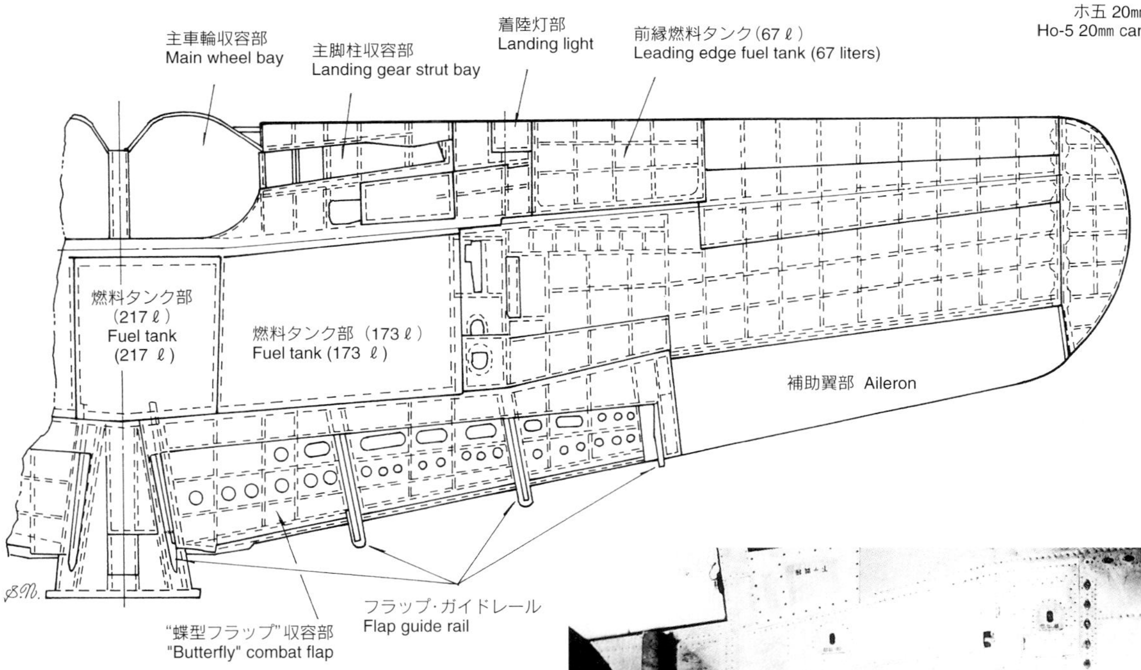

▼左下写真と同じ部分の外板を外し、内部の骨組みをみたところ。丸い肉抜き穴のあいたリブと、細い縦通材がわかる。画面上方は、収納状態の"蝶型フラップ"。

The same section as the photo at left seen with the outer panels removed. The ribs, with holes opened to lighten them, and the small lateral strengtheners are shown well. The "butterfly" combat flaps are visible in the stored position at the top of the photo.

▶左主翼下面の中央部付近。画面下が前縁で、左下の切り欠き部は前縁燃料タンクの収納スペース。中央に、落下増槽兼爆弾の懸吊架、その右にホ五 20mm機関砲の保弾子、空薬莢排出孔がある。右上の大きなボルトで止められたパネルが、燃料タンクの着脱用。

The center portion of the left wing, seen from the underside. The leading edge is at the bottom of the photo, and the cutout to the lower-left is where the leading edge fuel tank was installed. Bomb and external tank racks were mounted centrally, with the ejection ports for shells and used clips from the Ho-5 to the right of that. The panel to the upper right secured with relatively large bolts is to allow removal of the wing fuel tanks.

Flap structure (left side, units: mm)
フラップ骨組み図（左翼）(寸法単位mm)

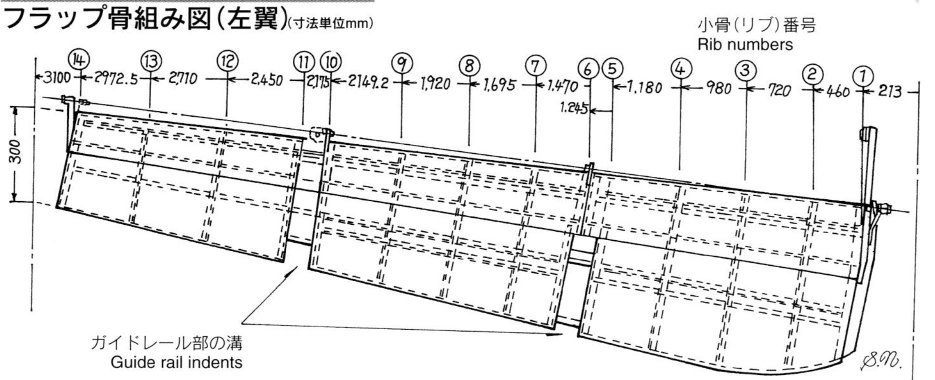

▲"蝶型フラップ"と呼ばれ、キ43、44にも装備された中島戦闘機独特のフラップは、もともと翼面荷重の高いキ44に、相応の旋回能力を与えるために考案されたものだ。機構的にはファウラー・フラップなのだが、後方への滑動範囲が外側より内側の方が大きく、そのようすが蝶の羽の動きを連想させることから、蝶型と、命名された。もっとも、実際には蝶の羽はこれとは逆に外側の方が大きく動くわけであるが……。この蝶型フラップは油圧により作動し、通常の離着陸時のほか、空中戦の際には操縦桿のボタンにより、15度まで下げられる。

The "butterfly" combat flaps installed on the Ki-43 and 44 were of a design that was unique to Nakajima and designed to boost the maneuverability of aircraft with relatively high wing-loading such as the Ki-44. Structurally, they are essentially the same as fowler flaps, though the movement to rearward is greater on the inboard side than the outboard one, creating a shape somewhat reminiscent of the wing of a butterfly, hence the name. Of course, an actual butterfly's wing drops more on the outboard side than the inboard. The flaps were hydraulically powered and could be raised and lowered during air combat up to 15° via a button on the control stick, as well as for normal take-off and landing use.

Flap guide rail mechanism
フラップ･ガイドレール取り付け要領

① 主翼小骨
② 主翼上面縦通材
③ 主翼上面外板
④ 主翼後縁
⑤ 先端保護木片
⑥ ガイドレール

① Wing rib
② Wing upper side horizontal strengtheners
③ Wing upper skin
④ Trailing edge
⑤ Wooden tip protector
⑥ Guide rail

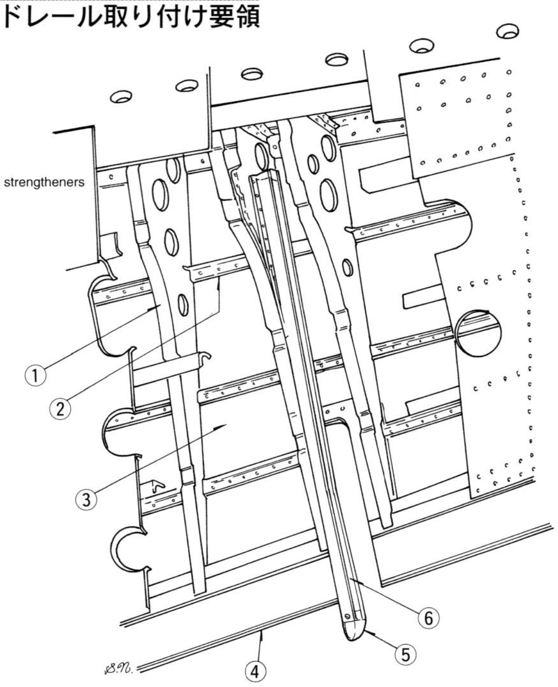

Aileron structure (units: mm)
補助翼骨組み図 (寸法単位mm)

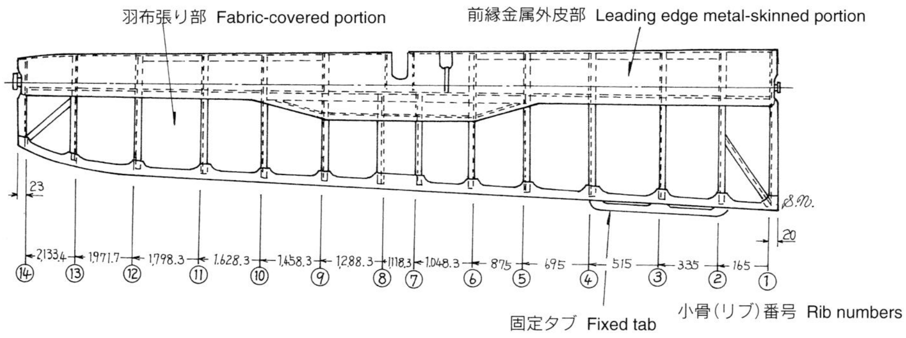

Wing fuel tank installation (right wing, top view)
主翼内燃料タンク収容部上面図（右翼）

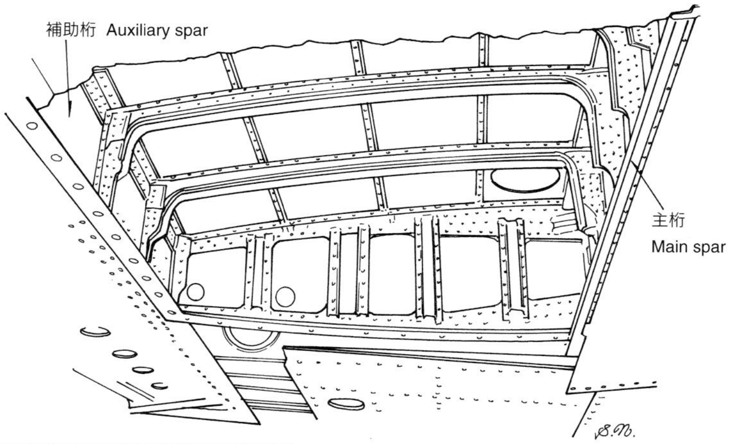

Horizontal stabilizer structure (units: mm)
水平安定板骨組み図 (寸法単位mm)

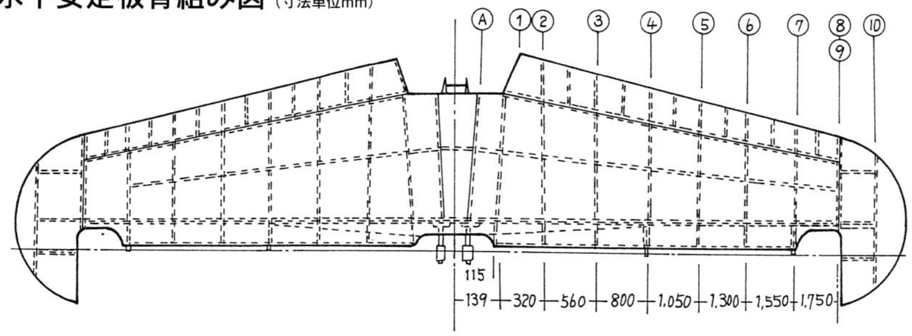

▲キ84は、エンジン出力の大きさもあってか、各動翼の最適な構造、面積を決定するのに、中島のスタッフもかなり苦労したようだ。増加試作機における、方向舵の骨組みに関しても、キ84甲の初期生産機とそれ以降では、リブの本機が異なり、後者ではより多くなっている。上、および下に示した補助翼、昇降舵骨組図は、初期のものを示す。

Having been given an unusually large amount of engine power to work with, the Ki-84's design team apparently had difficulty in deciding on the best structure and size for the aircraft's flight control surfaces. This is evident from the variations in the shape of the plane's rudder among the several prototypes. The internal structure of the plane's ailerons and elevator also differ between early Ki-84 production models and later planes, with the latter having more ribs. The diagrams above and right of the elevator and ailerons portray the early version.

Elevator structure (units: mm)
昇降舵骨組み図 (寸法単位mm)

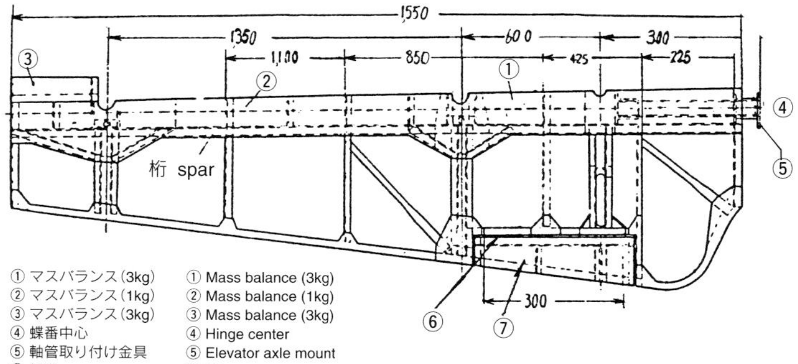

① マスバランス(3kg)
② マスバランス(1kg)
③ マスバランス(3kg)
④ 蝶番中心
⑤ 軸管取り付け金具
⑥ 蝶番
⑦ 修正タブ

① Mass balance (3kg)
② Mass balance (1kg)
③ Mass balance (3kg)
④ Hinge center
⑤ Elevator axle mount
⑥ Hinge
⑦ Correction tab

Nakajima "Ha 45-21" engine
中島『ハ45-21』エンジン

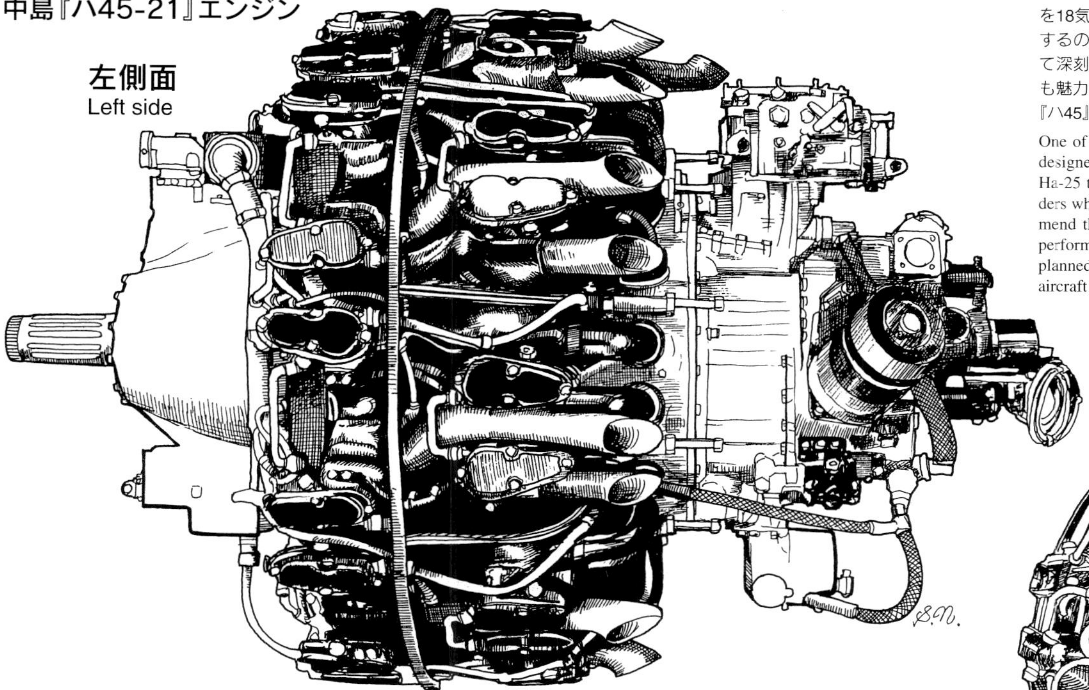

◀キ84開発の端緒となり、かつその障害を左右したのは、中島の自社製『ハ45』(海軍名称『誉』)エンジンであった。成功を収めた複列14気筒の『ハ25』~『ハ115』系(海軍名称『栄』)を18気筒化し、サイズ、重量の増加を最小限におさえながら、2,000馬力のパワーを実現するのがねらいだった。しかし、多分に理想ばかりを追究した設計は、実用段階になって深刻な不調やトラブルを発生させ、結果的に信頼性が低く計画出力も出ない、なんとも魅力のないエンジンになってしまった。キ84にとって、他の諸処の悪条件にもまして、『ハ45』エンジンの不調は、誤算の最たるものだった。

One of the biggest factors affecting the ultimate performance of the Hayate was the Nakajima-designed Ha-45 powerplant. Nakajima attempted to modify the company's highly-successful Ha-25 to Ha-115 series of 14-cylinder engines to 2,000hp of power by adding four more cylinders while keeping size and weight increases to a minimum. However, there was little to recommend the resulting engine. Burdened by over-ambitious design, it was plagued by numerous performance and maintenance problems, not the least of which was that it failed to deliver the planned-for amount of power. The failure of the Ha-45 was no doubt the biggest burden on an aircraft already besieged by other troubles.

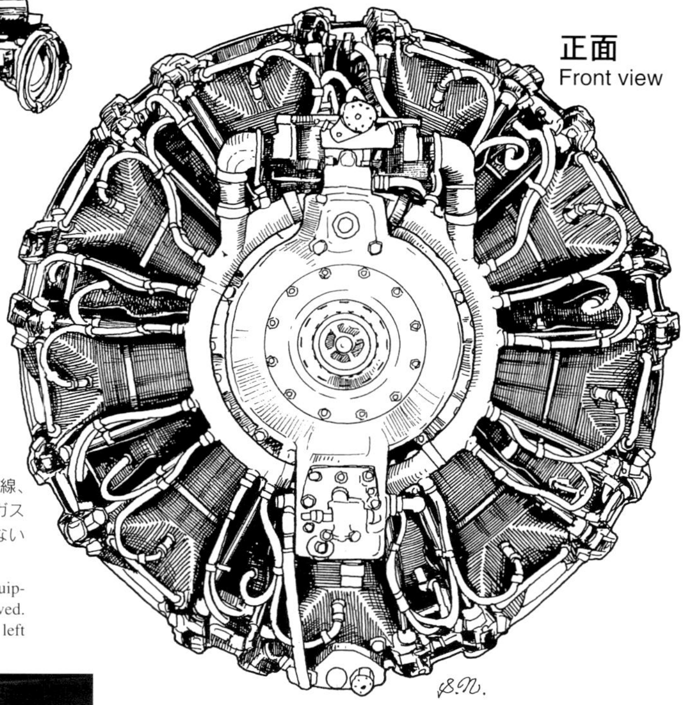

▼カウリング・パネルをはずし、『ハ45-21』エンジンの左側をみる。各シリンダー間隔はせまく、排気管、コード類、バッフルプレートなどが隙間を埋め、みるからにきゅうくつな印象をあたえる。

A view of the right side of Ha 45-21 engine, seen through removed cowl access panels. Note the extremely cramped conditions among the various lines, pipes and even the cylinders themselves.

▼◀ カウリング後方のパネルをはずし、エンジン補器類、配線、配管、および潤滑油タンクをみる。ホー〇三機関砲の爆発筒ガス抜き穴のディテールなどは、復原機では把握することのできない部分でもある。左写真は右側、右写真は左側を示す。

A view of the exhaust pipes, wiring, oil tank and other auxiliary equipment at the rear of the engine, seen with the rear cowl panels removed. This area is not historically accurate on the restored aircraft. The left photo is of the right side, while the right photo is of the left.

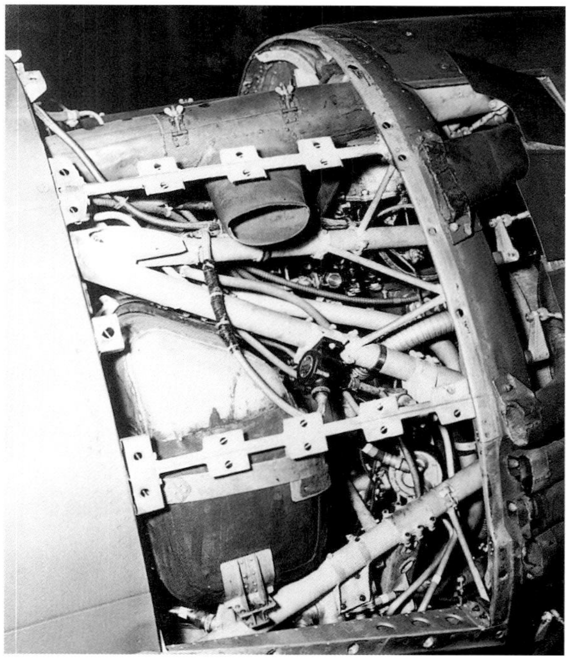

Cockpit Forward View
操縦席正面

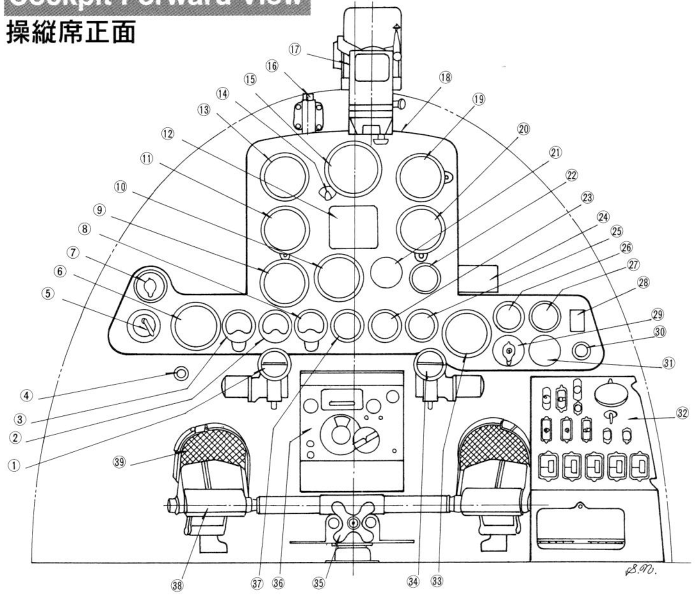

① 主翼内ホ五 20㎜機関砲装填コック
② 排気温度計
③ シリンダー温度計
④ 超過与圧操作レバー
⑤ 一号点火開閉器
⑥ 降着装置標示器
⑦ プロペラ管制器
⑧ 一式潤滑油温度計
⑨ 百式回転計
⑩ 二式吸入圧力(ブースト)計
⑪ 三式速度計
⑫ 九八式羅針儀
⑬ 九八式旋回指示器
⑭ 人工水準器コック
⑮ 九八式人工水準器(水平儀)
⑯ 室内通風(換気)装置レバー
⑰ 三式光像式射撃照準器
⑱ 主計器板
⑲ 九七式昇降計
⑳ 九七式高度計
㉑ 百式航空時計
㉒ 酸素流量計
㉓ 九五式燃料圧力計
㉔ 自差表
㉕ 一式燃料計
㉖ 二速与圧器用油圧計
㉗ タンク加圧用燃料圧力計
㉘ 旋回指示器調整弁
㉙ 油量計切換開閉器
㉚ 速度計排雨器
㉛ タンク加圧用切換開閉器
㉜ 配電盤
㉝ 九五式油量計
㉞ ホー〇三 12.7㎜機関砲用装填コック
㉟ 方向舵踏棒前後調整ハンドル
㊱ 飛三号無線機用受信器
㊲ 九四式潤滑油圧力計
㊳ 方向舵ペダル
㊴ ブレーキ・ペダル

① Wing-mounted Ho-5 20㎜ cannon arming cock
② Exhaust temperature gauge
③ Cylinder temperature gauge
④ Overboost control lever
⑤ Model 1 ignition switch
⑥ Landing gear position indicator
⑦ Propeller speed control
⑧ Type 1 oil temperature gauge
⑨ Type 100 tachometer
⑩ Type 2 boost pressure gauge
⑪ Type 3 speed indicator
⑫ Type 98 compass
⑬ Type 98 turn indicator
⑭ Artificial horizon cock
⑮ Type 98 artificial horizon
⑯ Cockpit air flow control lever
⑰ Type 3 reflecting gunsight
⑱ Main instrument panel
⑲ Type 97 vertical speed indicator
⑳ Type 97 altimeter
㉑ Type 100 aeronautic clock
㉒ Oxygen flow gauge
㉓ Type 95 fuel gauge
㉔ Data table
㉕ Type 1 fuel gauge
㉖ Two-speed compressor oil pressure gauge
㉗ Tank pressurization fuel pressure gauge
㉘ Turn indicator adjustment
㉙ Oil gauge selector
㉚ Air speed indicator rain remover
㉛ Tank pressurization selector
㉜ Electrical box
㉝ Type 95 oil guage
㉞ Ho-103 12.7㎜ MG loading cock
㉟ Rudder pedal position adjustment handle
㊱ "Hi" Mk.3 wireless receiver
㊲ Type 94 oil pressure gauge
㊳ Rudder pedal
㊴ Brake pedal

三式速度計
Type 3 speed indicator

九七式高度計
Type 97 altimeter

九八式旋回指示器
Type 98 turn indicator

九七式昇降計
Type 97 vertical speed indicator

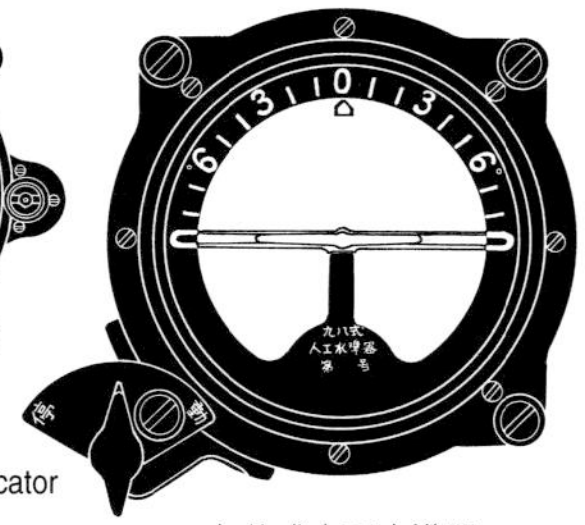

九八式人工水準器
Type 98 artificial horizon

二式吸入圧力(ブースト)計
Type 2 boost pressure gauge

一式燃料計
Type 1 fuel gauge

水メタノール圧力計
Water/methanol pressure gauge

燃料タンク加圧計
Fuel tank pressure gauge

シリンダー温度計
Cylinder temperature gauge

百式回転計
Type 100 tachometer

九五式燃圧計
Type 95 fuel pressure gauge

燃料計切換スイッチ
Fuel gauge selector switch

排気温度計
Exhaust temperature gauge

▶キ84甲の操縦者正面計器板。機首上部ホー〇三 12.7㎜機関砲の砲尾をクリアするため、逆"T"字形をしている。

The main instrument panel of the Ki-84 Ko. The panel is shaped in an upside-down 'T' configuration to give clearance to the ends of the nose-mounted Ho-103 12.7mm machine guns.

◀キ84乙の操縦室正面計器板。機首上部兵装がホ五 20㎜ 機関砲に換装されたための措置と思われるが、計器板が単純な長方形に変わった。

The main instrument panel of the Ki-84 Otsu. Note the redesign of the panel into a standard rectangle due to the change of nose armament to the Ho-5 20mm cannon.

Cockpit left side
操縦席左側

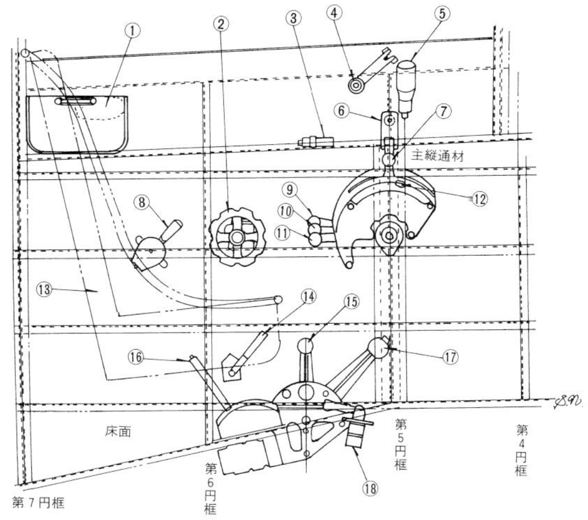

① 送受話器格納袋
② 昇降修整舵操作輪
③ 移動灯
④ 風防開閉レバー
⑤ 紫外線灯
⑥ スロットル・レバー
⑦ プロペラ・ピッチ手動操作レバー
⑧ 無線機遠隔操作器
⑨ 過給器零速切換レバー
⑩ 過給器自動高空弁切換レバー
⑪ 過給器二速切換レバー
⑫ 高度弁調節レバー
⑬ 座席
⑭ 座席高低調節レバー
⑮ 尾輪固定レバー
⑯ フラップ操作レバー
⑰ 降着装置操作レバー
⑱ 油圧用二方コック

① Radio mic/speaker storage pouch
② Elevator trim tab control
③ Movable light
④ Canopy open lever
⑤ Ultraviolet light
⑥ Throttle lever
⑦ Manual propeller pitch control lever
⑧ Radio remote control
⑨ Supercharger zero-speed selector lever
⑩ Supercharger automatic high-altitude valve selector lever
⑪ Supercharger two-speed selector lever
⑫ High altitude valve adjustment lever
⑬ Seat
⑭ Seat height adjustment lever
⑮ Tail wheel lock control
⑯ Flap control
⑰ Landing gear control
⑱ Oil pressure two-way cock

Cockpit right side
操縦席右側

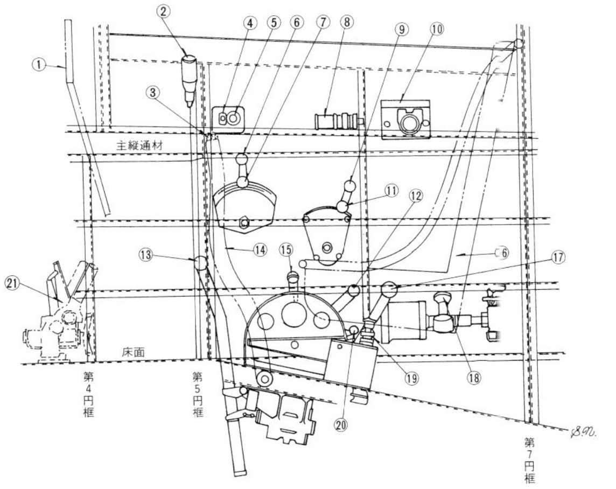

① 主計器板
② 紫外線灯
③ 機関砲発射ボタン
④ 電源開閉器
⑤ 起動用押ボタン
⑥ 潤滑油冷却器シャッター操作レバー
⑦ カウルフラップ操作レバー
⑧ 座席灯
⑨ 防塵装置操作レバー
⑩ 計数器（写真銃用）
⑪ 吸入空気暖気レバー
⑫ 五方コック操作レバー
⑬ 油圧装置手動ポンプ・レバー
⑭ 操縦桿
⑮ 三方コック操作レバー
⑯ 座席
⑰ 燃料ポンプ手動レバー
⑱ 酸素調整器自動切換レバー
⑲ 燃料注射ポンプ
⑳ 四方コック操作レバー
㉑ 方向舵／ブレーキ・ペダル

① Main instrument panel
② Ultraviolet light
③ Cannon firing switch
④ Main power switch
⑤ Starter switch
⑥ Oil cooler shutter control
⑦ Cowl flap control lever
⑧ Seat light
⑨ Dust filter control lever
⑩ Gun camera film counter
⑪ Air warmer control lever
⑫ 5-way cock control lever
⑬ Manual oil pressure pump lever
⑭ Control stick
⑮ 3-way cock control lever
⑯ Seat
⑰ Manual fuel pump lever
⑱ Automatic oxygen adjuster selector
⑲ Injection pump
⑳ 4-way cock control lever
㉑ Rudder/brake pedals

▲（上）キ84甲の操縦室内左側。画面右手手前の操縦桿頂部に注目。（下）キ84甲の操縦室内右側。米軍接収後のもので、画面右上の無線機関係操作ボックスはオリジナルと異なるものが取り付けられている。

Above : The left side of the cockpit of the Ki-84 Ko. Note the top of the control stick in the right foreground. **Below :** The right side of the cockpit of the Ki-84. This photo was taken after the plane was captured by U.S. forces. As a result, the radio control box in the upper right of the photo is no longer original equipment.

Control stick
操縦桿

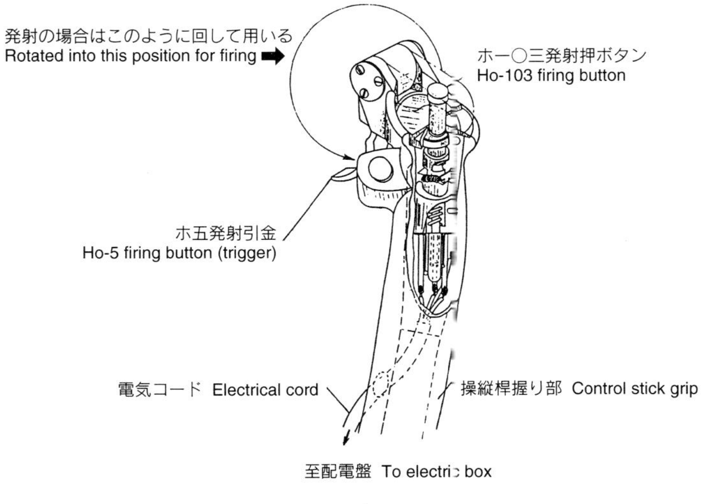

Type 3 reflecting gunsight
三式光像式射撃照準器

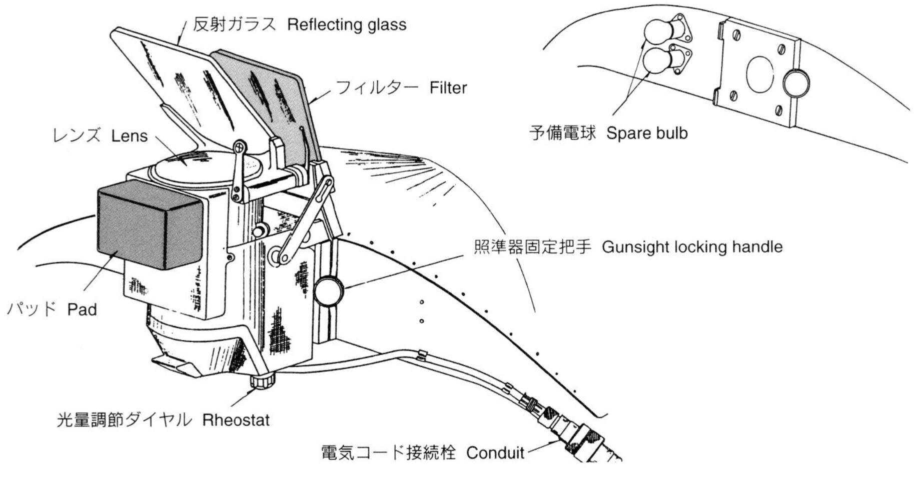

Canopy structure
風防構造図

風防骨格一般
Canopy Construction Detail

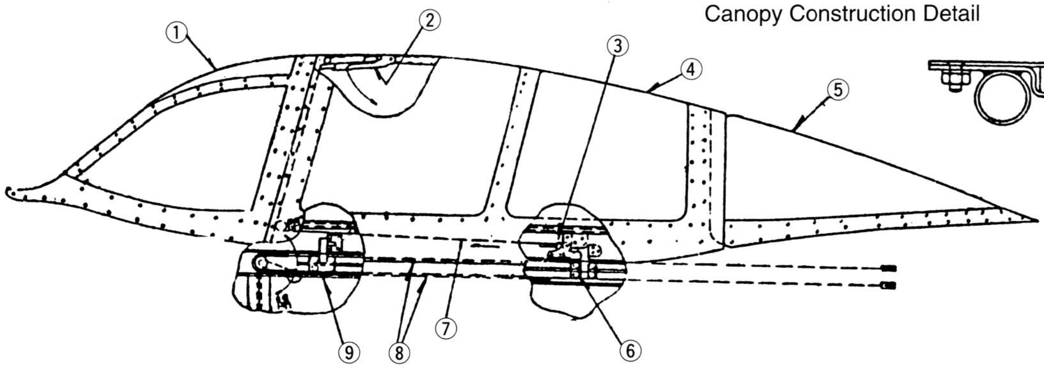

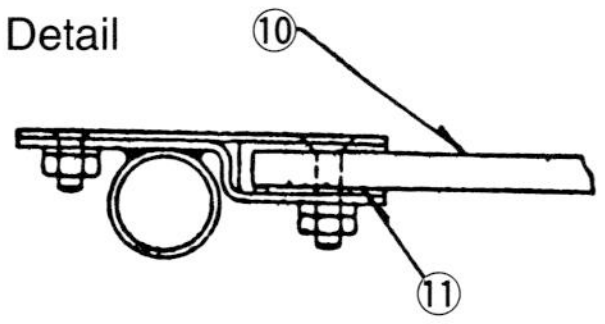

可動風防装置
Canopy open/close mechanism

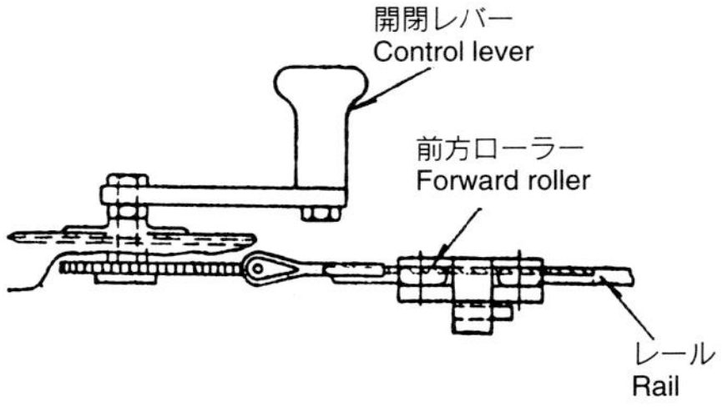

① 前部風防
② 非常時飛散装置ハンドル
③ 飛散装置
④ 可動風防
⑤ 後部風防
⑥ 後方ローラー
⑦ 飛散装置操作索
⑧ 風防開閉装置操作索と鎖
⑨ 前方ローラー
⑩ プレキシガラス
⑪ 緩衝ゴム

① Windscreen
② Emergency release lever
③ Canopy release mechanism
④ Movable canopy
⑤ Rear canopy
⑥ Rear roller
⑦ Release mechanism control cable
⑧ Canopy open/close control cable
⑨ Forward roller
⑩ Plexiglass
⑪ Shock-absorbing rubber

Armor plate installation
防弾鋼板取り付け要領

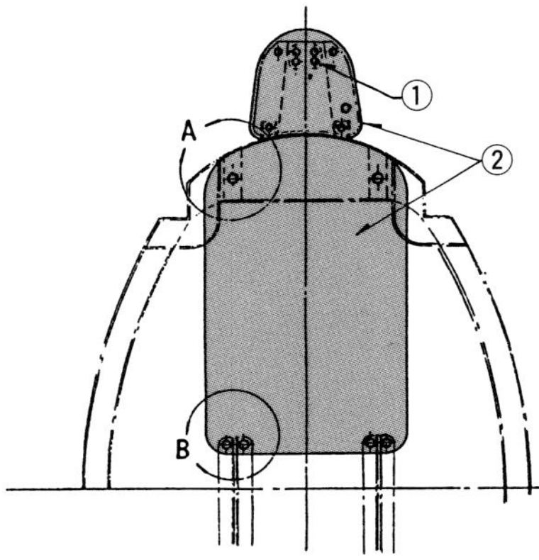

A部詳細
Section A detail

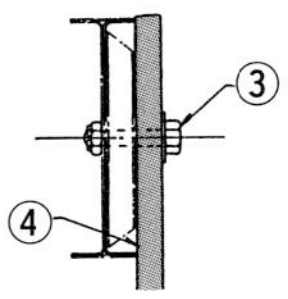

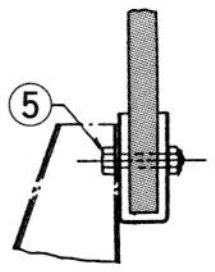

B部詳細
Section B detail

① 6mm一号ボルト
② 13mm厚防弾鋼板
③ 8mm一号ボルト
④ 防弾鋼板
⑤ 6mm一号ボルト

① 6mm #1 bolts
② 13mm armor plate
③ 8mm #1 bolt
④ Armor plate
⑤ 6mm #1 bolt

Pilot's seat
座席

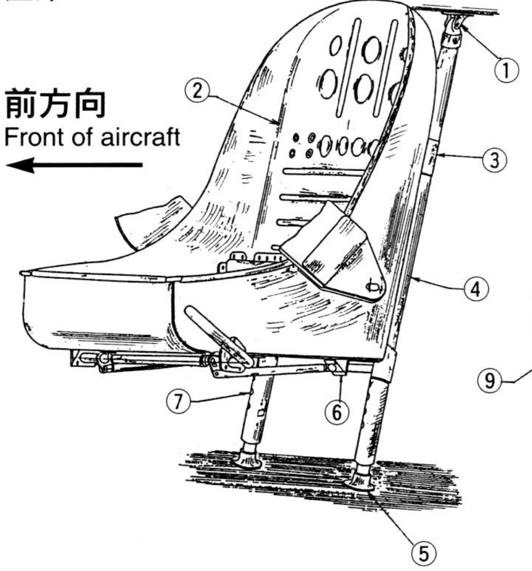

後方よりみる
Rear view

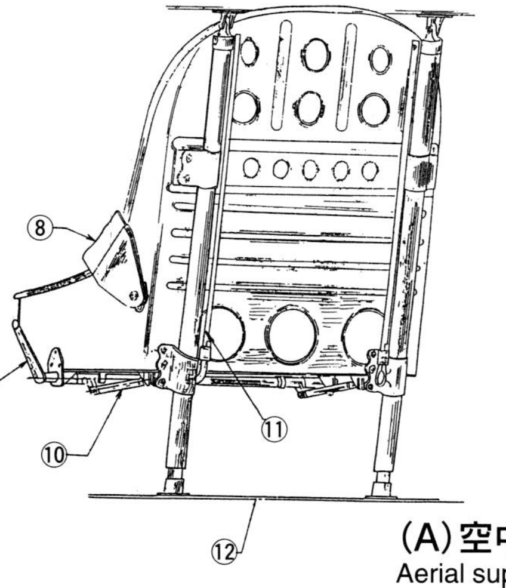

① 支柱取り付け金具
② 座席本体
③ 摺動金具
④ 支柱
⑤ 座席取り付け金具
⑥ 止め金具
⑦ 高さ調節用穴
⑧ 安全バンド
⑨ 高低調節レバー
⑩ 連動管
⑪ 緩衝ゴム
⑫ 床

① Support pipe mount
② Seat
③ Slide attachment
④ Main support pipe
⑤ Seat mounts
⑥ Position lock
⑦ Height adjustment holes
⑧ Safety band
⑨ Seat leg mount
⑩ Linked adjustment pipe
⑪ Shock-absorbing rubber
⑫ Floor

Aerial installation
アンテナ空中線取り付け要領

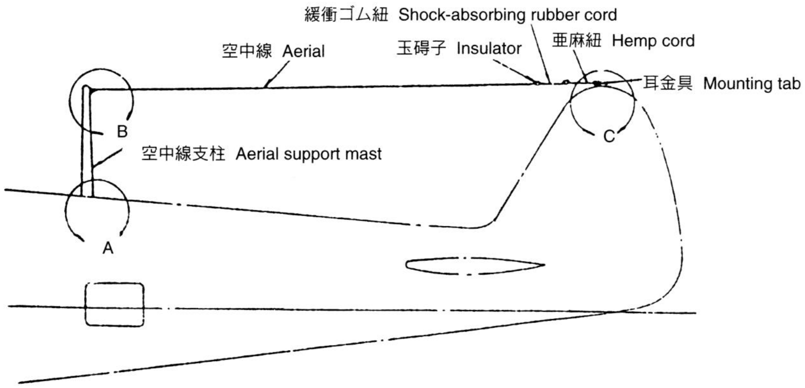

(A)空中線支持取り付け
Aerial support mast attachment

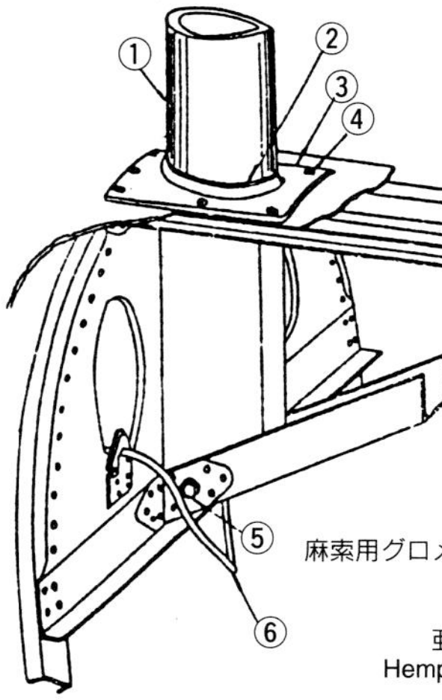

① 空中線支柱
② ゴム間座
③ 抑え板
④ 4mm丸小ネジ
⑤ 8mmボルト
⑥ 引き込み線

① Aerial support mast
② Rubber mount
③ Securing plate
④ 4mm screw
⑤ 8mm bolt
⑥ Retracting wire

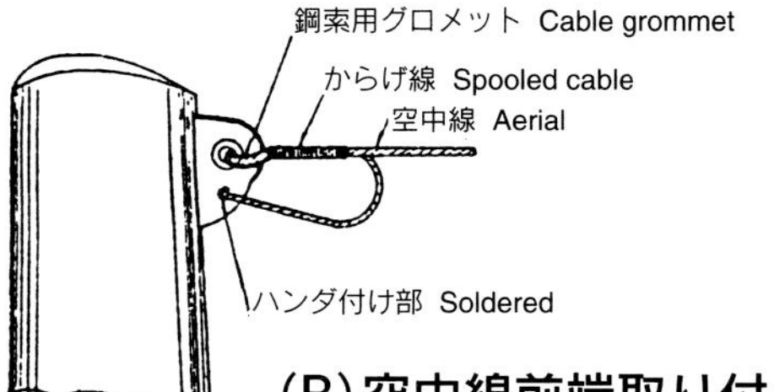

(B)空中線前端取り付け
Forward aerial attachment

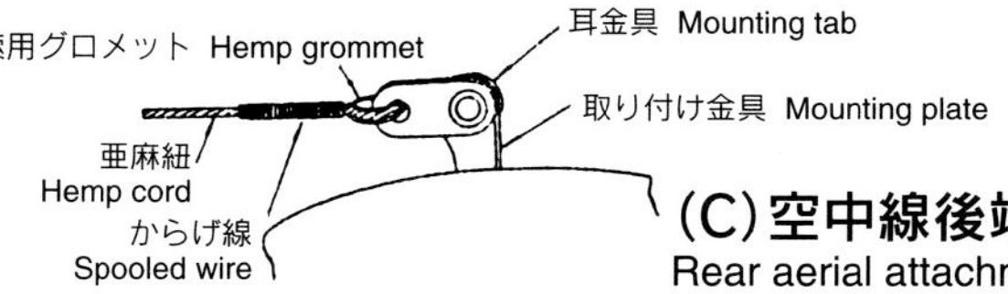

(C)空中線後端取り付け
Rear aerial attachment

Main gear structure
主脚の構成

① オレオ脚柱（クロームモリブデン鋼製）
② 回転主軸
③ 注油栓
④ 片持脚柱（ジュラルミン鋼物）
⑤ トルクアーム
⑥ 車輪（650×170㎜、内圧4.5気圧）
⑦ 車軸
⑧ 押金具
⑨ 懸吊金具
⑩ 防塵覆
⑪ 脚位置表示棒

① Oleo strut (chrome-molybdenum)
② Rotation axle
③ Strut oil fill
④ Single-side wheel mount (duralumin)
⑤ Torque arm
⑥ Main tire (650 x 170㎜, 4.5 atmos. pressure)
⑦ Axle
⑧ Wheel mounting plate
⑨ Lock pin
⑩ Dust cover
⑪ Gear position indicator rod

試作機～増加機のホイール・ハブ内側

Wheel hub interior of prototypes

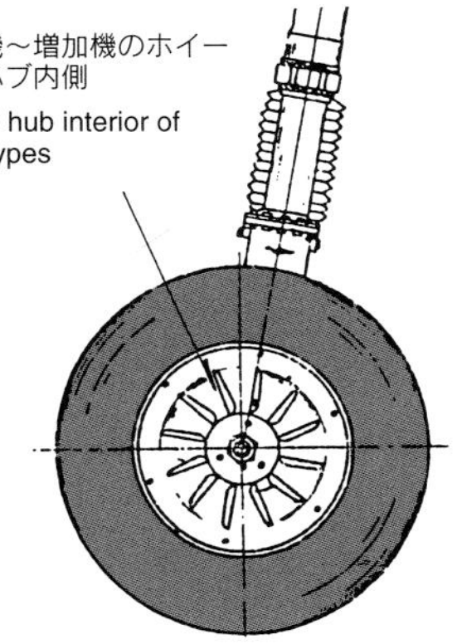

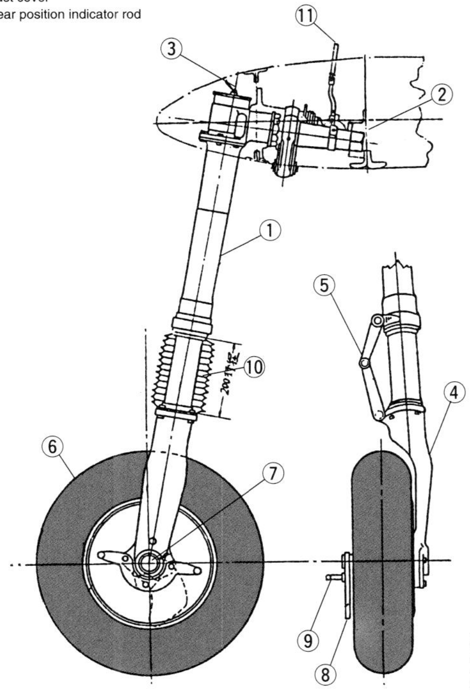

Main gear cover
主脚柱カバー

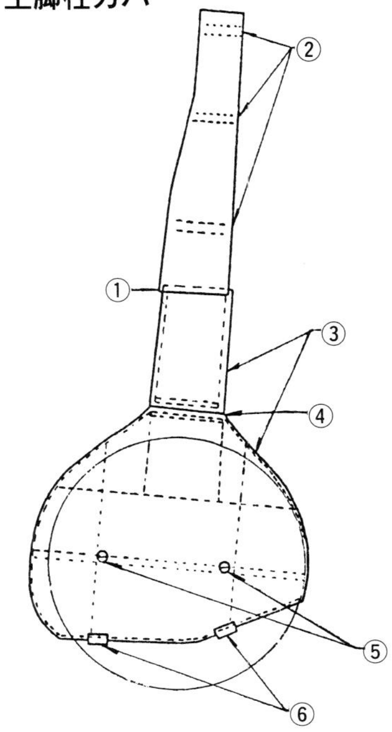

① 内面のカバーと外面のカバーが干渉する場合は取り付けボルトで調整
② 取り付け金具（調整可能）
③ 脚上げが困難なときにこの部分が当たる場合あり
④ ブレーキ調整時はここから取りはずす
⑤ 取り付けボルト
⑥ 制限板

① Cover overlap adjustment bolt
② Adjustable cover mounts
③ Gear cover and bay sometimes fail to align properly here
④ Removed during brake maintenance
⑤ Attachment bolt
⑥ Limiter panel

▲キ84甲の右主脚。シンプルな構成で、生産性は良いようだ。タイヤは650×170㎜サイズで、直径は零戦のそれより50㎜大きい。

The right main gear of the Ki-84 Ko. The simple design made for easy manufacturing. The 650 x 170mm tire was 50mm larger in diameter than that of the Zero.

Main wheel cover
主車輪カバー

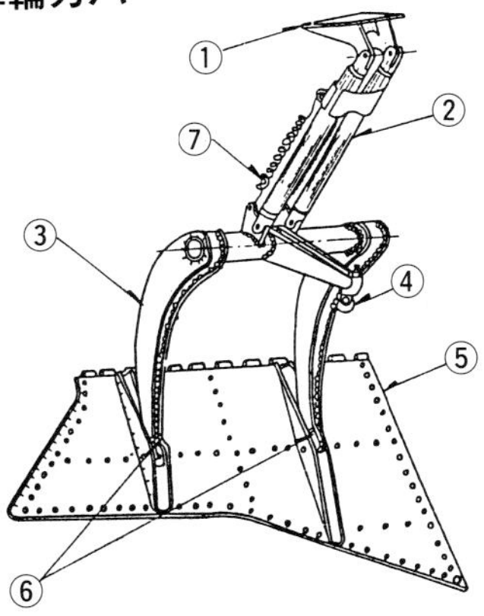

① 取り付け金具
② 上部操作ロッド
③ 下部操作アーム
④ ローラー
⑤ 車輪カバー
⑥ 調整ボルト
⑦ バネ

① Mount
② Upper retraction arm
③ Lower retraction arm
④ Roller
⑤ Wheel cover
⑥ Adjustment bolt
⑦ Spring

◀キ84甲の尾脚。タイヤは200×75㎜サイズのソリッド・ゴム製。収納孔から泥、ホコリが入らぬよう、ゴム布製のカバーが付けられている。

The tail gear of the Ki-84 Ko. The tire is solid rubber, measuring 200 x 75mm. A rubber cover is installed to help prevent mud and debris from entering the gear bay.

Tail gear structure
尾脚の構成

尾脚下げ状態
Tail gear lowered position

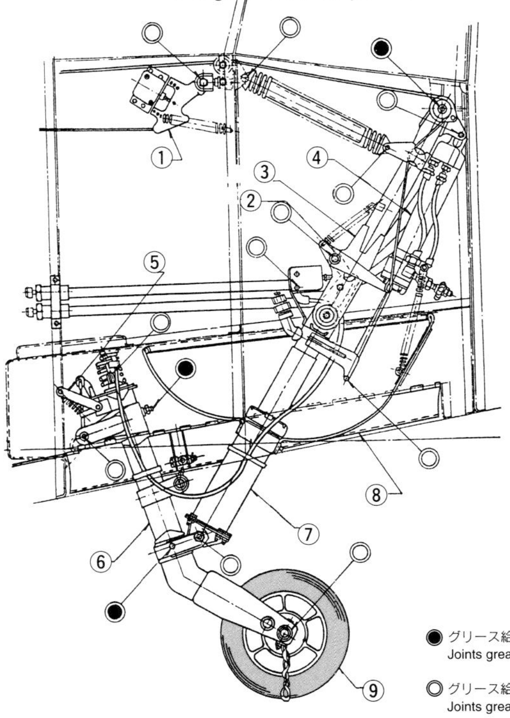

尾脚上げ状態
Tail gear raised position

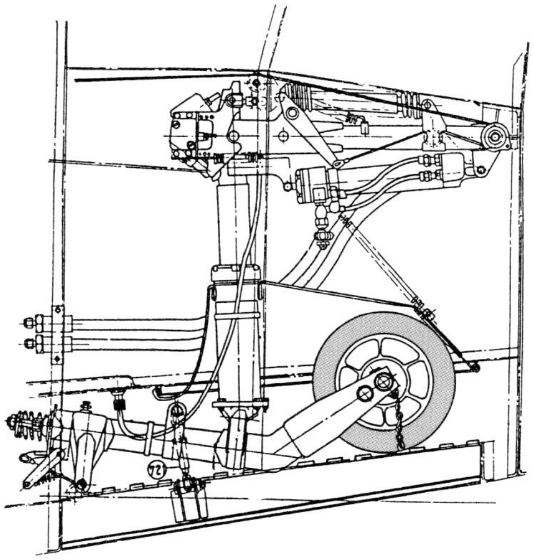

● グリース給油
Joints greased externally

○ グリース給油金具を有す
Joints greased internally

① アップロック装置
② ダウンロック装置
③ 上部支柱
④ 作動筒
⑤ 尾輪固定装置
⑥ 尾輪支柱
⑦ オレオ
⑧ 収容扉
⑨ 尾輪（200×75㎜）

① Raised position lock
② Lowered position lock
③ Upper support strut
④ Retraction arm
⑤ Tail wheel lock
⑥ Tail gear strut
⑦ Oleo
⑧ Tail gear bay door
⑨ Tire (200 x 75㎜)

Armament Installation (units: mm) 射撃兵装位置関係図（寸法単位mm）

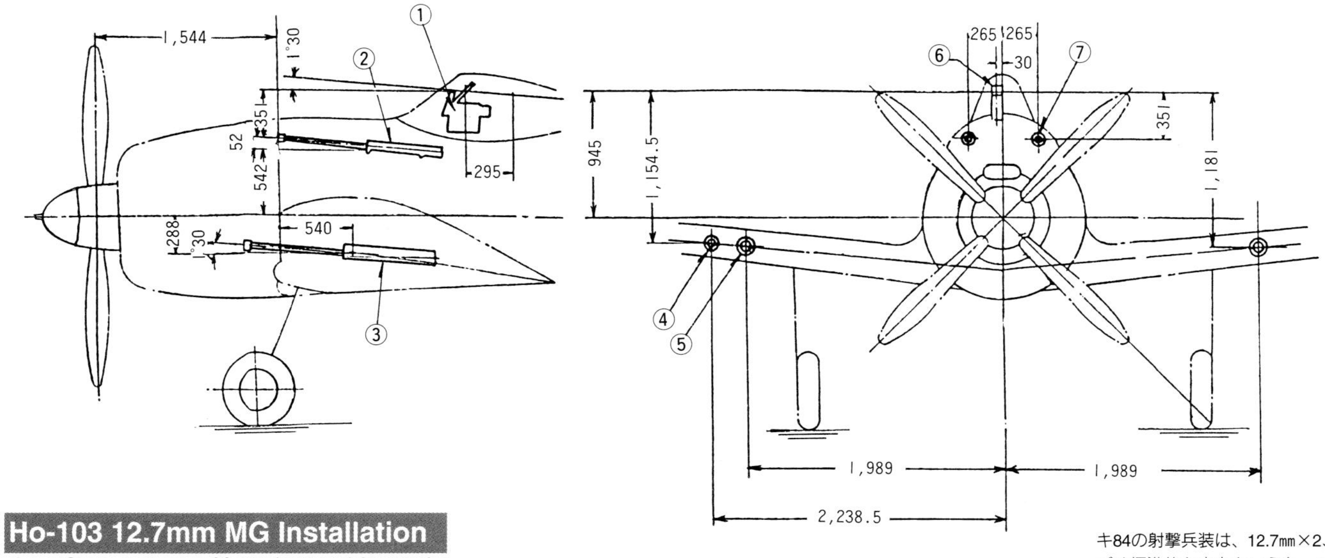

① 三式射撃照準器
② ホー〇三
③ ホ五
④ 三式射撃照準器
⑤ 射撃監査写真機（ガンカメラ／オプション）
⑥ 三式射撃照準器
⑦ ホー〇三

① Type 3 reflecting gunsight
② Ho-103 MG
③ Ho-5 cannon
④ Type 3 reflecting gunsight
⑤ Gun camera (optional)
⑥ Type 3 reflecting gunsight
⑦ Ho-103 MG

Ho-103 12.7mm MG Installation
ホー〇三 12.7㎜機関砲取り付け要領

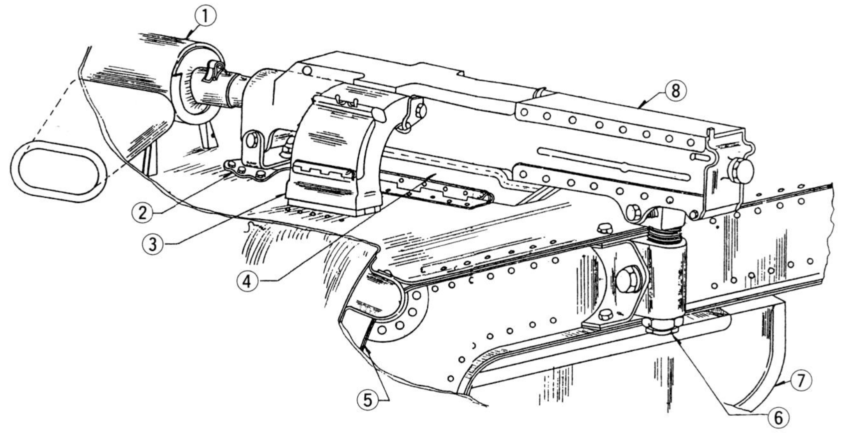

① 冷却筒
② 前方取り付け金具
③ 上部給弾口
④ 上部薬莢排出筒
⑤ 胴体第3円框（フレーム）
⑥ 後方取り付け金具
⑦ 保弾子収容箱
⑧ ホー〇三機関砲

① Cooling tube
② Forward mount
③ Upper shell feed opening
④ Upper casing ejection port
⑤ Fuselage rib #3
⑥ Rear mount
⑦ Used clip storage
⑧ Ho-103 MG

キ84の射撃兵装は、12.7㎜×2、20㎜×2で、設計着手当時の単発戦闘機としては、ごく標準的な火力といえた。それでも、陸軍の単発戦闘機としては、計画当初から20㎜機関砲を予定した最初の機体だったわけで、海軍の零戦に比較して実に5年の遅れがある。このことだけみても、太平洋戦争中期までの陸軍戦闘機が、連合軍機に対し、常に火力不足のハンディキャップを負いながら戦わざるを得なかったことがよくわかる。海軍についても事情は同じだが、陸軍の航空機銃／機関砲もまた、その大半が外国製品のコピー、もしくはライセンス品、およびそれらの改良品で占められてた。キ43、44、61の主力火器ともなったホー〇三 一式12.7㎜機関砲は、アメリカのコルト・ブローニングM2 12.7㎜機銃のコピーで、太平洋戦争の開戦直後から、やっとキ43に本格装備され始めた。信じ難いことだが、コルト・ブローニングM2がなければ、日本陸軍戦闘機は7.7㎜機銃で太平洋戦争を戦わねばならなくなるところだった。

The armament of the Ki-84 was designed from the outset to be of the typical Japanese 12.7mm x 2, 20mm x 2 configuration. However, it's interesting to note that this was the first Japanese Army aircraft to have 20mm cannon installed from the time of the plane's initial conception, a full five years behind the Navy's Zero. This fact alone strongly suggests the large firepower handicap that IJA fighters suffered when facing much more heavily-armed American fighters. What's more, though this was the same for the Navy, Army aircraft armaments were primarily copies, modifications or license-produced versions of foreign guns. The Ho-103 installed on the Ki-43, 44 and 61 was a copy of the U.S. Colt-Browning M2 12.7mm MG and didn't see use on the Ki-43 until shortly after the war began. Though hard to believe, if the M2 hadn't existed, IJA fighters would have had to try to fight the war armed with only 7.7mm MGs.

Ho-103 Firing control system
ホー〇三装塡発射装置

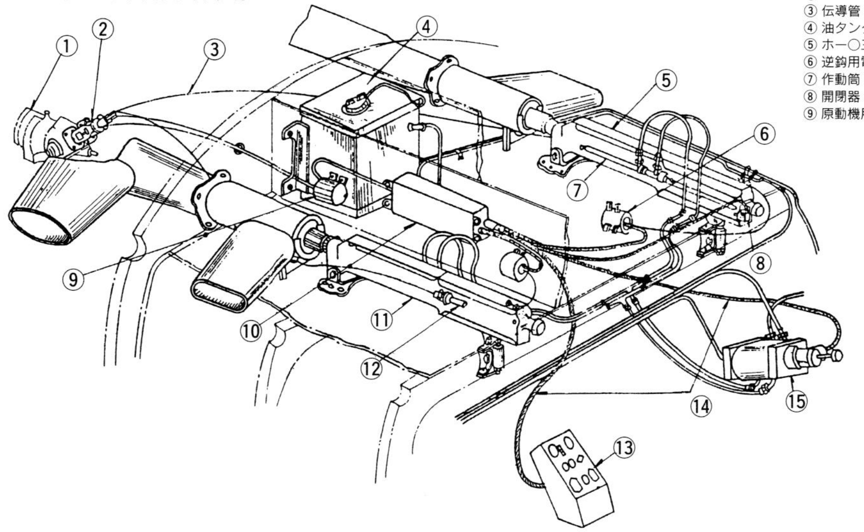

① 自動進角装置
② プロペラ同調装置原動機
③ 伝導管
④ 油タンク
⑤ ホー〇三（甲）
⑥ 逆鈎用電磁器
⑦ 作動筒
⑧ 開閉器
⑨ 原動機用電磁器
⑩ 前部接続筐
⑪ ホー〇三（乙）
⑫ 撃発機
⑬ 配電盤
⑭ ケーブル
⑮ 装塡コック

① Propeller speed automatic synchro device
② Propeller synchronization device
③ Wiring
④ Oil tank
⑤ Ho-103 (Ko)
⑥ Bolt lock synchro switch
⑦ Actuator tube
⑧ Trigger control
⑨ Synchro device switch
⑩ Forward connection box
⑪ Ho-103 (Otsu)
⑫ Firing control switch
⑬ Power supply box
⑭ Wiring
⑮ Loading cock

Ho-103 12.7mm Machine Gun
ホー〇三 12.7㎜機関砲

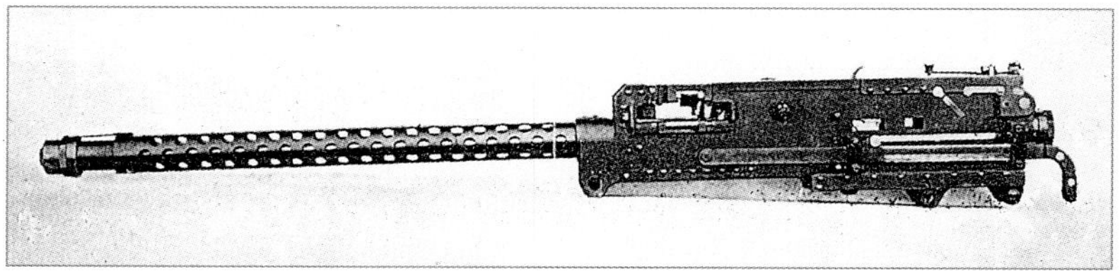

▲ホー〇三の性能は、もとが信頼性抜群の優秀品なので、実用性は申し分なく、炸裂弾を使えるようにしたぶん、オリジナルより弾丸そのものの威力が増していた。

The Ho-103, being a copy of a superb weapon, proved itself to be excellent armament. The modifications to allow it to fire exploding shells actually gave it slightly more punch than the original.

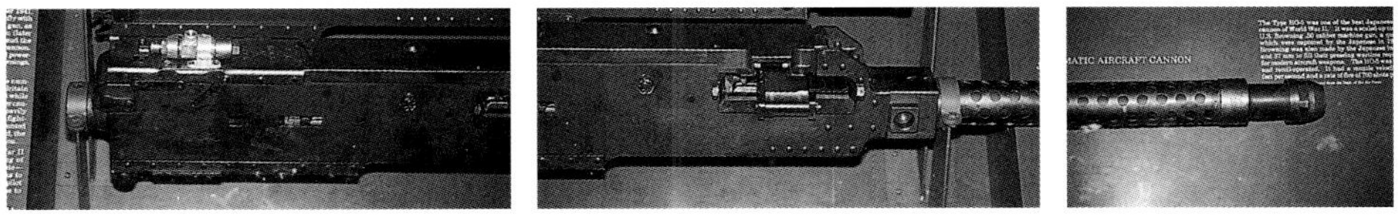

▲アメリカのNASM本館に展示されているホ五 20㎜機関砲。その外形からして、コルト・ブローニングM2ベースの拡大版であることが、ひと目でわかる。

A Ho-5 20mm cannon on display at the National Air & Space Museum of the Smithsonian Institute in Washington, D.C. The silhouette quickly gives away that the gun is an enlarged version of the M2.

▶キ84甲の左主翼前縁から突き出した、ホ五 20mm機関砲。機軸に対し1°30′の上向き角をつけて取り付けられている。

The muzzle of a Ho-5 20mm jutting out of the leading edge of the wing of a Ki-84 Ko. The guns were mounted angled up 1°30' from the aircraft's centerline.

Ho-5 20mm cannon installation (diagram shows left side)

ホ五 20㎜機関砲取り付け要領 (図は左翼を示す)

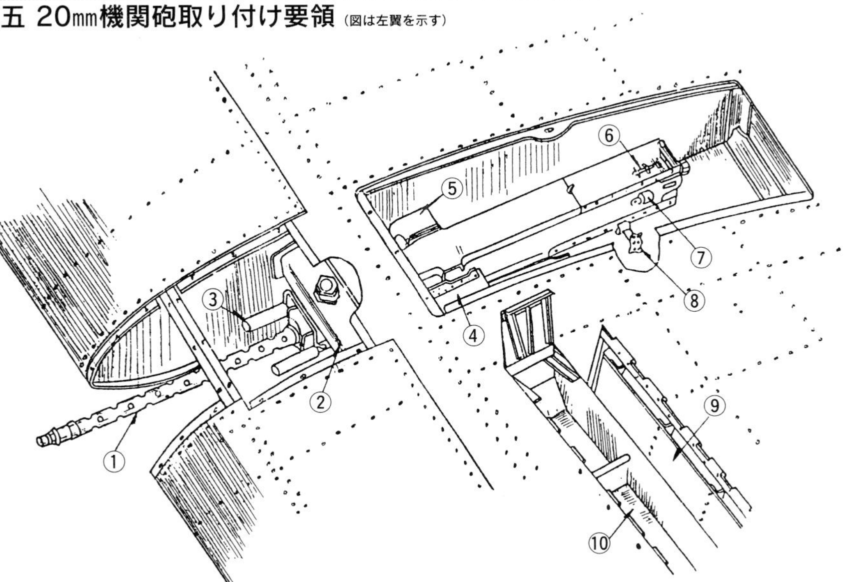

① ホ五機関砲
② 前方支持金具
③ 駐退器
④ 給弾口
⑤ 保弾子受
⑥ 逆鈎作動筒
⑦ 引金用電磁器
⑧ 後方支持金具
⑨ 弾薬箱蓋
⑩ 弾薬箱

① Ho-5 cannon
② Cannon mount
③ Barrel extention spring
④ Shell feed opening
⑤ Used clip chute
⑥ Bolt lock control
⑦ Trigger switch
⑧ Cannon holder
⑨ Wing magazine access hatch
⑩ Wing magazine

ホ五もまた、20㎜機関砲不在に慌てた陸軍が、ホ一〇三をスケール・アップして急ぎ採用したものである。当然、構造はコルト・ブローニングM2のそれを踏襲している。ホ五はキ84の増加試作機から優先的に装備され、昭和19年以降キ45改、キ61などにも装備された。性能的には、海軍のエリコン系九九式二号 20mm機銃をしのいだ。キ84甲は、左右翼内とも各150発の弾丸を携行した。

The Ho-5 20mm cannon was a rapidly-designed, scaled-up version of the Ho-103 implemented to give the IJA's fighters heavy firepower for the first time. As such, the basic design borrows heavily from the Colt-Browning M2. Installed first on the Ki-84 beginning with the prototype, the weapon also saw use in the Ki-45 kai, Ki-61 and others beginning in 1944. It proved itself a superior weapon to the Type 99 Mk.2 20mm cannon that saw use in Navy aircraft. Each of the two guns could carry 150 rounds in wing magazines.

Shell feed system (front view)

弾丸搭載要領（正面図）

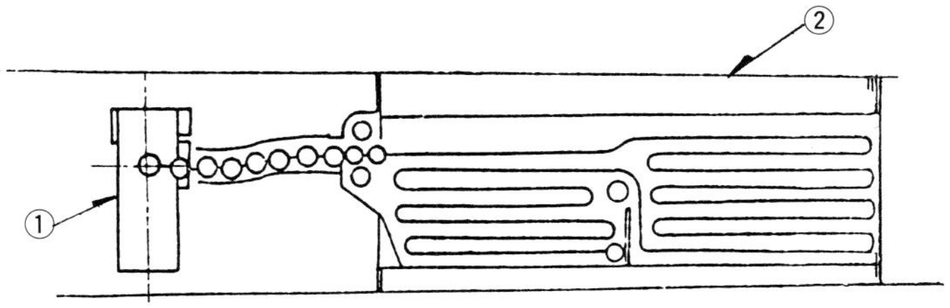

① ホ五機関砲
② 弾薬箱

① Ho-5 cannon
② Wing magazine

Underwing drop tank / bomb carrier

爆弾懸吊具（落下増槽懸吊具も兼ねる）

① 主翼外板
② 主翼10番小骨
③ 丸小ネジ用ストップナット
④ 主翼10番小骨
⑤ 信管翼
⑥ 安全針
⑦ 前方弾抑え
⑧ 電磁器
⑨ 丸小ネジ
⑩ 回り止め
⑪ 調整ネジ部
⑫ 安全針
⑬ 安全針止め金具

① Outboard wing surface
② Small wing rib #10
③ Stop nuts
④ Small wing rib #10
⑤ Signal flare tube
⑥ Safety pin
⑦ Forward sway stopper
⑧ Generator
⑨ Round screw
⑩ Rotation stopper
⑪ Adjustment screw
⑫ Safety pin
⑬ Safety pin stopper

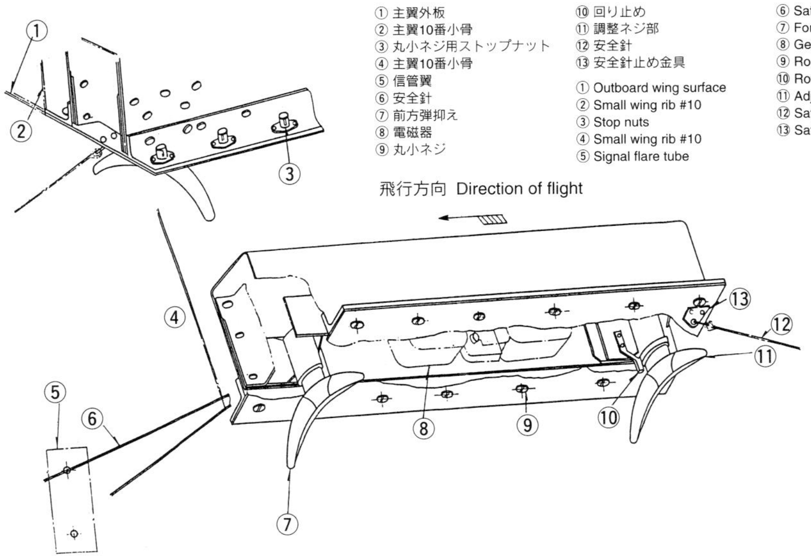

▼キ84甲の左主翼下面に懸吊された、統一型200ℓ入落下増槽。増槽は木製、竹製2種あったが、写真のものは前者。このラックが爆弾と共用で、その際は250kgまで懸吊可能。4個ある三日月形の振れ止め金具のうち、内側の小さいもの2個は、爆弾懸吊の際に使う。

The standard type 200-liter drop tank seen slung under the left wing of the Ki-84. The tanks were constructed of either wood or bamboo, the one in the photo being the former. The rack could be used for bombs as well, able to mount up to 250kg weapons. The smaller, inner two of the four sway stoppers visible here were used when bombs were mounted.

Type 95 "Hi" Mk.III Radio Installation
九五式飛三号無線機および接続要領

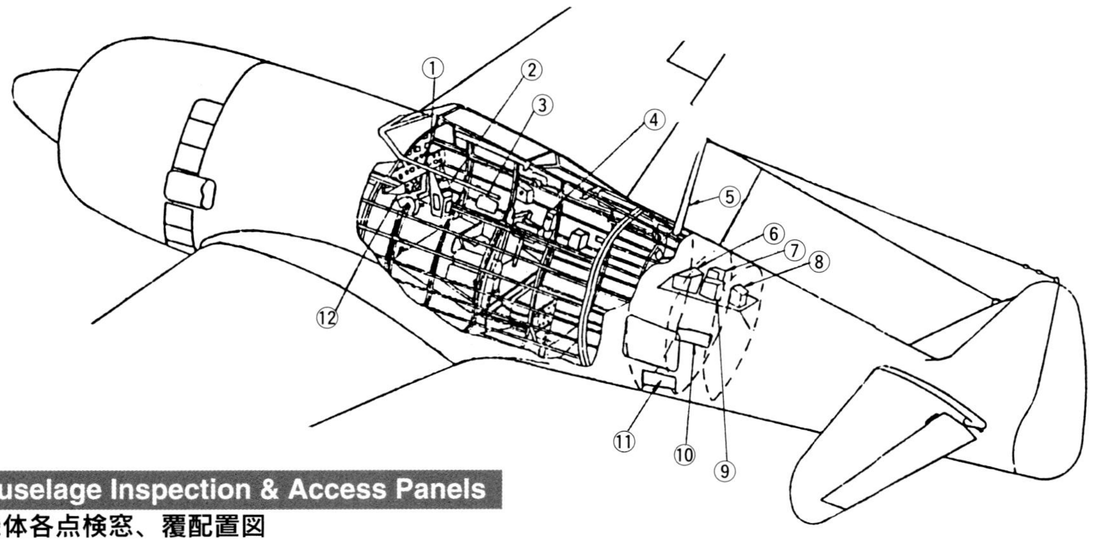

① 電鍵
② 配電盤
③ 送信話器格納袋
④ 後部接続筐
⑤ 空中線装置
⑥ 変調器
⑦ 空中線同調器
⑧ 受信器
⑨ 送信器
⑩ 接続函／受話器／送信器
⑪ 直流変圧器
⑫ 遠操器

① Morse key
② Circuit board
③ Mic/speaker storage pouch
④ Rear connection box
⑤ Aerial support mast
⑥ Tuner
⑦ Aerial tuner
⑧ Receiver
⑨ Transmitter
⑩ Connection box/Handset/transmitter
⑪ DC transformer
⑫ Remote control

キ84が搭載した無線器は、昭和14年（1939年）制式採用の、九九式飛三号と称する小型機用だった。零戦二一型から二二型を通して搭載された、海軍の九六式空一号よりは新しいものの、その信頼性がどれほどのものだったか、当事者たちの回想がほとんど聞かれないのでわからない。なにしろ、昭和18年制式採用の海軍三式空一号でさえ、同20年2月以降に撃墜した米海軍戦闘機F4Uの無線機を検分したことで、アースの適切な処理方法がやっと判り、それからどうにか役に立つようになったといわれているくらいだから、陸軍機用無線機も、メーカーこそ違え、同じような状況にあったことは間違いない。なお、キ84の一部は、特別装備として飛五号（長距離用？）を搭載した。

The radio in the Ki-84 was the Type 99 "Hi" Mk.III, originally put into service beginning in 1939. Though a slightly newer unit that the Type 96 "Ku" Mk.I that was in service in the Zero Model 21 and 22, there is little information remaining about how useful a piece of equipment it was (incidentally, the Type 96 was widely regarded by Navy pilots as completely useless, and many removed them from their aircraft in order to save weight, in defiance of regulations). In any event, it wasn't until Japanese technicians studied the radio grounding method of an F4U Corsair that was shot down in 1945 that the Navy's 1943-design Type 3 "Ku" Mk.I began to be a truly reliable aircraft communications system. As such, it is hard to imagine that the Type 99 could have been any better previous to that. Some Ki-84s mounted the "Hi" Mk.V (believed to be a longer-range radio) instead.

Fuselage Inspection & Access Panels
機体各点検窓、覆配置図

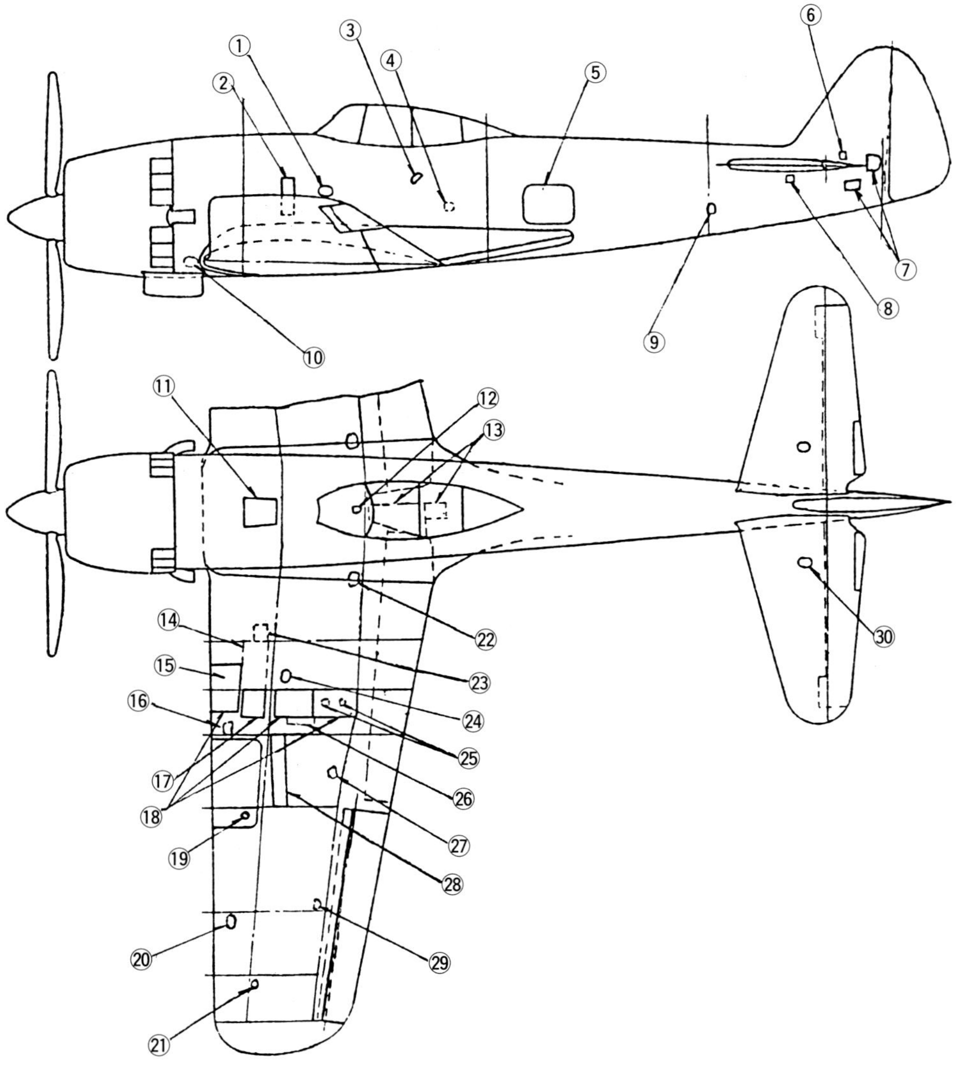

① 胴体下タンク燃料積入口蓋
② 胴体内弾倉室覆
③ 手掛
④ 電源接続筐点検窓蓋
⑤ 作業口扉
⑥ 尾輪降着装置点検窓蓋(両側)
⑦ 尾輪降着装置点検覆
⑧ 水平安定板取り付け部点検窓蓋(両側)
⑨ かつぎ棒穴蓋(両側)
⑩ エンジン始動操作点検窓覆
⑪ メタノールタンク燃料積入口蓋および作動油タンク注積入口蓋
⑫ 胴体下タンク燃料排出蓋
⑬ 下翼中央索筒点検覆
⑭ 降着装置点検覆(左右)
⑮ 脚軸点検覆(左右)
⑯ 監査写真機点検窓蓋（右翼のみ)
⑰ 翼内砲駐退器点検覆(左右)
⑱ 翼内砲点検覆(左右)
⑲ 前縁タンク燃料積入口(左右)
⑳ 速度計点検窓覆
㉑ 繋留索点検窓覆(左右)
㉒ 翼内タンク燃料排出口蓋(左右)
㉓ 脚操点検覆(左右)
㉔ 翼内タンク燃料積入口蓋(左右)
㉕ 翼内砲点検窓蓋(左右)
㉖ 爆弾懸吊架点検窓覆(左右)
㉗ 補助翼第2槓桿点検窓覆(左右)
㉘ 翼内弾倉室覆(左右)
㉙ 補助翼第3槓桿点検窓覆(左右)
㉚ 修整舵鼓胴点検覆（左右)

① Lower fuselage fuel tank fill
② Fuselage magazine cover
③ Hand hold
④ External power connection
⑤ Access hatch
⑥ Tail gear inspection hatch (both sides)
⑦ Tail gear access hatches
⑧ Horizontal stabilizer mount inspection hatch (both sides)
⑨ Fuselage hoist opening (both sides)
⑩ Engine starter inspection window
⑪ Methanol tank/oil tank fill access hatch
⑫ Lower fuselage fuel tank drain hatch
⑬ Central wing towing hook cover
⑭ Landing gear inspection hatch (both sides)
⑮ Gear axle inspection hatch (both sides)
⑯ Camera inspection hatch present on right wing of recon. planes
⑰ Wing cannon breech inspection hatch (both sides)
⑱ Wing cannon inspection hatch (both sides)
⑲ Leading edge fuel tank fill
⑳ Air speed indicator inspection window cover
㉑ Towing hook cover (both sides)
㉒ Wing fuel tank drain cover (both sides)
㉓ Gear control inspection hatch (both sides)
㉔ Wing fuel tank fill (both sides)
㉕ Wing cannon inspection window cover (both sides)
㉖ Bomb rack inspection window cover (both sides)
㉗ Aileron #2 control rod inspection hatch (both sides)
㉘ Wing magazine access hatch (both sides)
㉙ Aileron #3 control rod inspection hatch(both sides)
㉚ Trim tab control bar inspection hatch (both sides)

四式戦闘機『疾風』各型変遷　Variations of the Ki-84

作図・解説／野原 茂
Illustrations & commentary by Shigeru Nohara

(1/48スケール)
(All diagrams are 1/48 scale)

キ84　第一次増加試作機　Ki-84 First Batch Supplementary Prototype, Early Type

① 砲口覆は長い
② 集合排気管
③ 頭部保護支柱は簡略タイプ
④ 翼下面懸吊装置はない
⑤ 主脚引込装置用リンク・アーム覆の形が異なる
⑥ 後縁が突出した方向舵

① Long muzzle cover
② Clustered exhaust pipes
③ Simple-type head rest
④ No underwing racks
⑤ Landing gear link arm cover shape differs
⑥ Vertical stabilizer trailing edge bulge pronounced

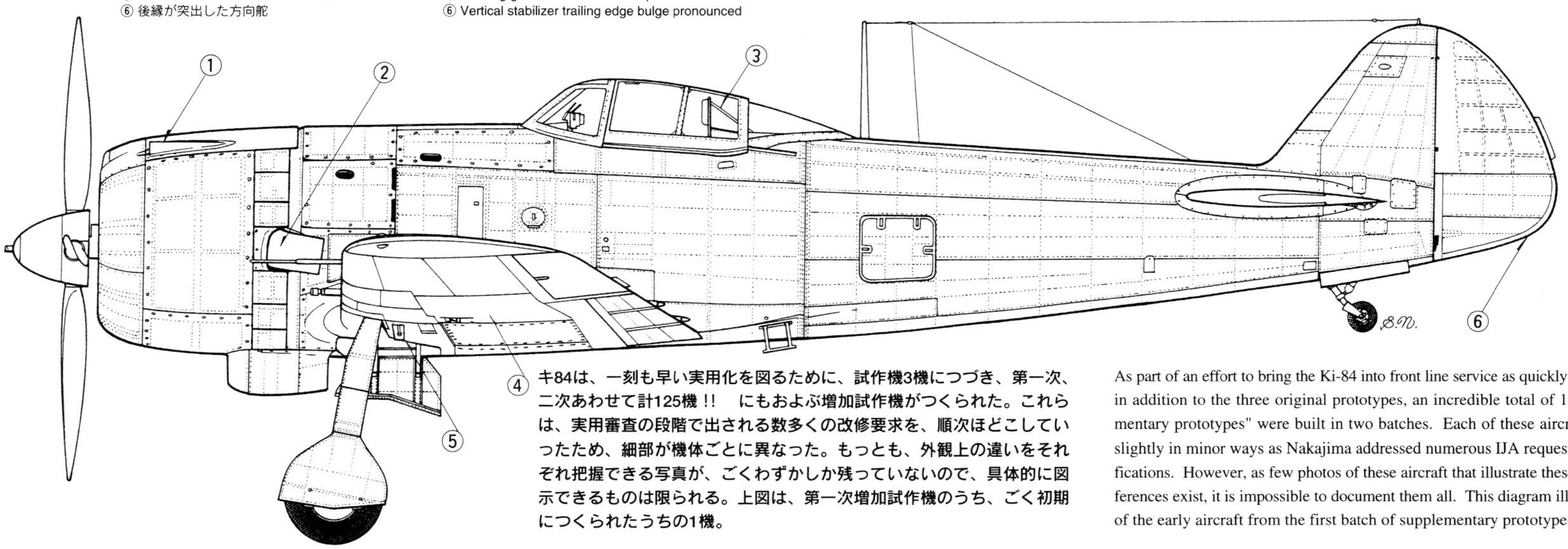

キ84は、一刻も早い実用化を図るために、試作機3機につづき、第一次、二次あわせて計125機！！　にもおよぶ増加試作機がつくられた。これらは、実用審査の段階で出される数多くの改修要求を、順次ほどこしていったため、細部が機体ごとに異なった。もっとも、外観上の違いをそれぞれ把握できる写真が、ごくわずかしか残っていないので、具体的に図示できるものは限られる。上図は、第一次増加試作機のうち、ごく初期につくられたうちの1機。

As part of an effort to bring the Ki-84 into front line service as quickly as possible, in addition to the three original prototypes, an incredible total of 125 "supplementary prototypes" were built in two batches. Each of these aircraft differed slightly in minor ways as Nakajima addressed numerous IJA requests for modifications. However, as few photos of these aircraft that illustrate these minor differences exist, it is impossible to document them all. This diagram illustrates one of the early aircraft from the first batch of supplementary prototypes.

キ84甲　量産機　Ki-84 Ko, Production Model

① プロペラ始動フック
② ラチエ式電気可変ピッチ定速プロペラ（直径3.05m）
③ ホー〇三 一式 12.7㎜機関砲
④ ホー〇三ガス抜口
⑤ ホー〇三弾倉筐
⑥ 光像式照準器
⑦ 翼内中央燃料タンク注入口ハッチ
⑧ 脚／フラップ警報ブザー
⑨ 手掛
⑩ 尾輪点検パネル
⑪ 尾灯
⑫ 方向舵トリム・タブ
⑬ 潤滑油冷却器覆
⑭ 主車輪（650×170㎜）
⑮ 主車輪カバー
⑯ 増槽懸吊時の振れ止め
⑰ 爆弾懸吊時の振れ止め
⑱ 足掛引き出しボタン
⑲ 引込式足掛
⑳ 引込式足掛
㉑ 足掛引き出しボタン
㉒ 胴体内部点検ハッチ
㉓ かつぎ棒差し込み穴蓋
㉔ 尾輪点検パネル
㉕ 尾輪点検パネル
㉖ 尾輪（200×75㎜）
㉗ 尾輪点検パネル

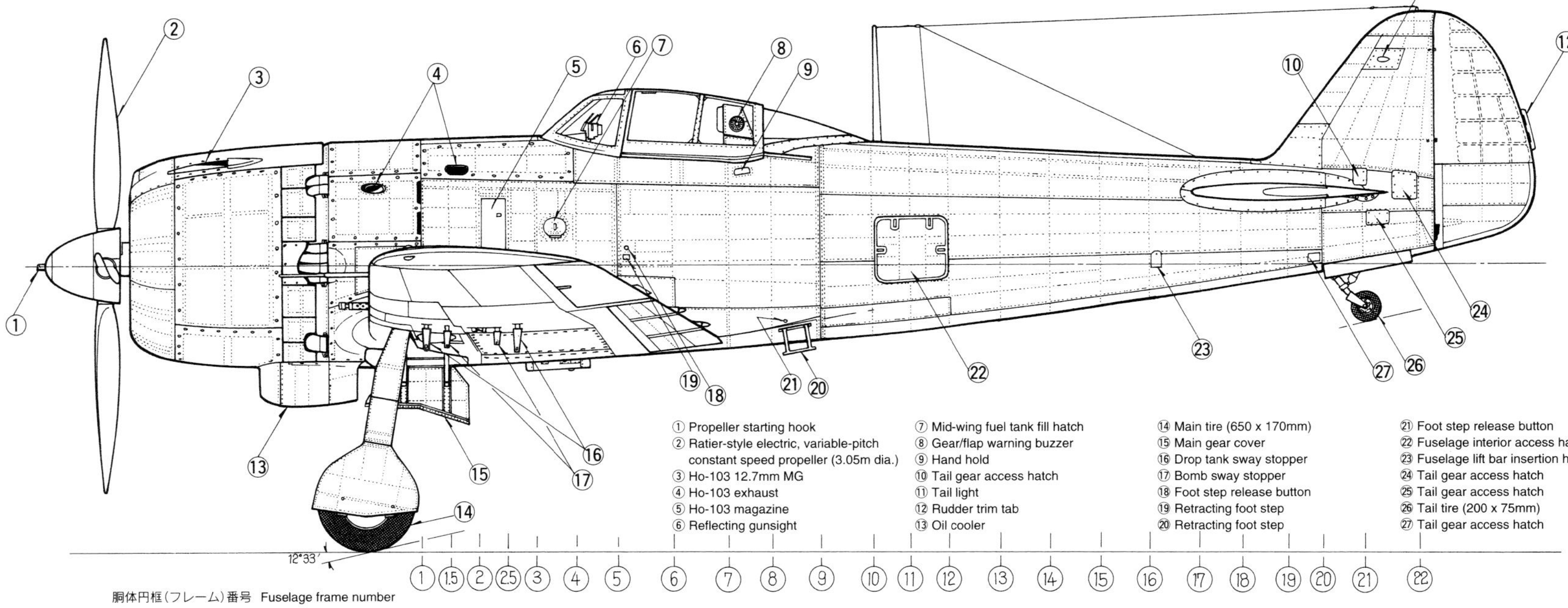

① Propeller starting hook
② Ratier-style electric, variable-pitch constant speed propeller (3.05m dia.)
③ Ho-103 12.7mm MG
④ Ho-103 exhaust
⑤ Ho-103 magazine
⑥ Reflecting gunsight
⑦ Mid-wing fuel tank fill hatch
⑧ Gear/flap warning buzzer
⑨ Hand hold
⑩ Tail gear access hatch
⑪ Tail light
⑫ Rudder trim tab
⑬ Oil cooler
⑭ Main tire (650 x 170mm)
⑮ Main gear cover
⑯ Drop tank sway stopper
⑰ Bomb sway stopper
⑱ Foot step release button
⑲ Retracting foot step
⑳ Retracting foot step
㉑ Foot step release button
㉒ Fuselage interior access hatch
㉓ Fuselage lift bar insertion hole
㉔ Tail gear access hatch
㉕ Tail gear access hatch
㉖ Tail tire (200 x 75mm)
㉗ Tail gear access hatch

胴体円框（フレーム）番号　Fuselage frame number

キ84甲　右側面図　Ki-84 Ko Right Side

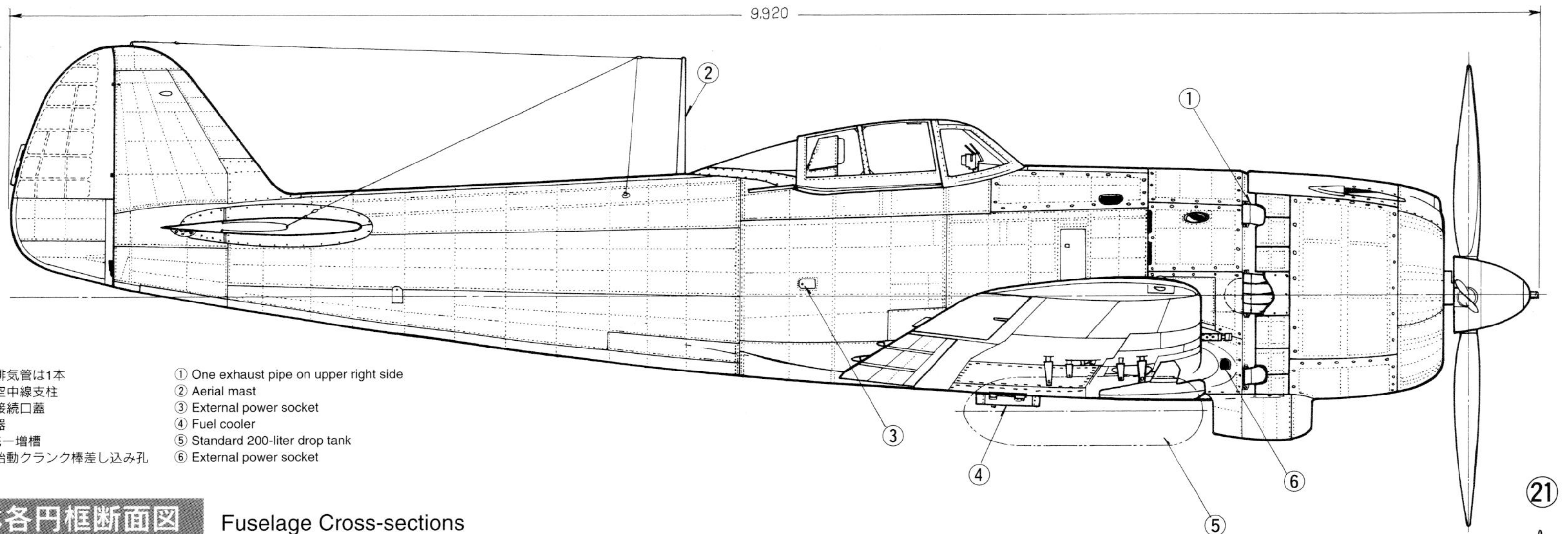

① 右側上方排気管は1本
② アンテナ空中線支柱
③ 外部電源接続口蓋
④ 燃料冷却器
⑤ 200 ℓ 入統一増槽
⑥ エンジン始動クランク棒差し込み孔

① One exhaust pipe on upper right side
② Aerial mast
③ External power socket
④ Fuel cooler
⑤ Standard 200-liter drop tank
⑥ External power socket

胴体各円框断面図　Fuselage Cross-sections

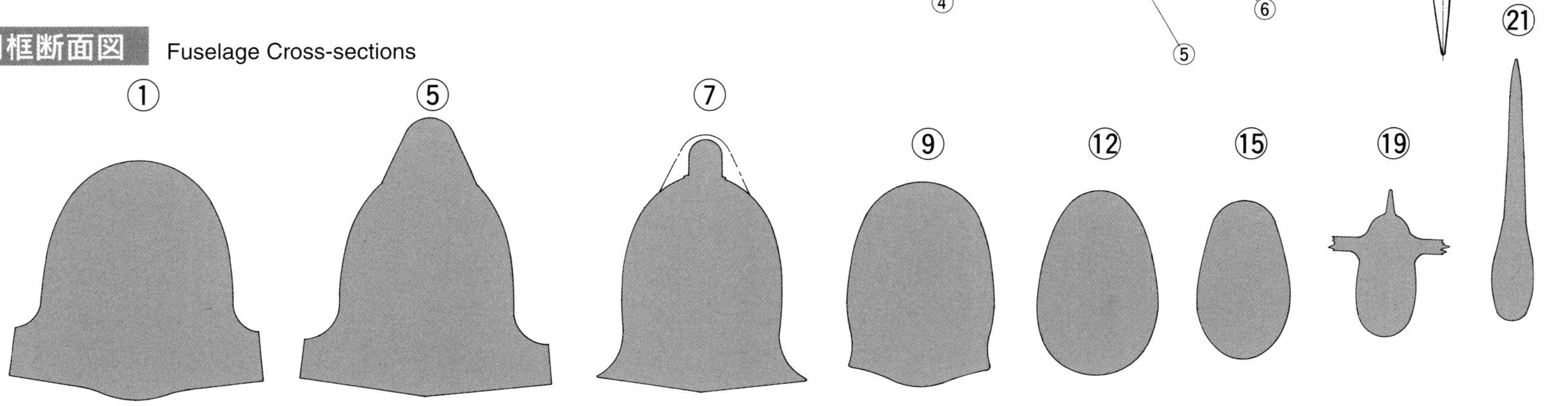

キ84甲　正面図　Ki-84 Ko Front View

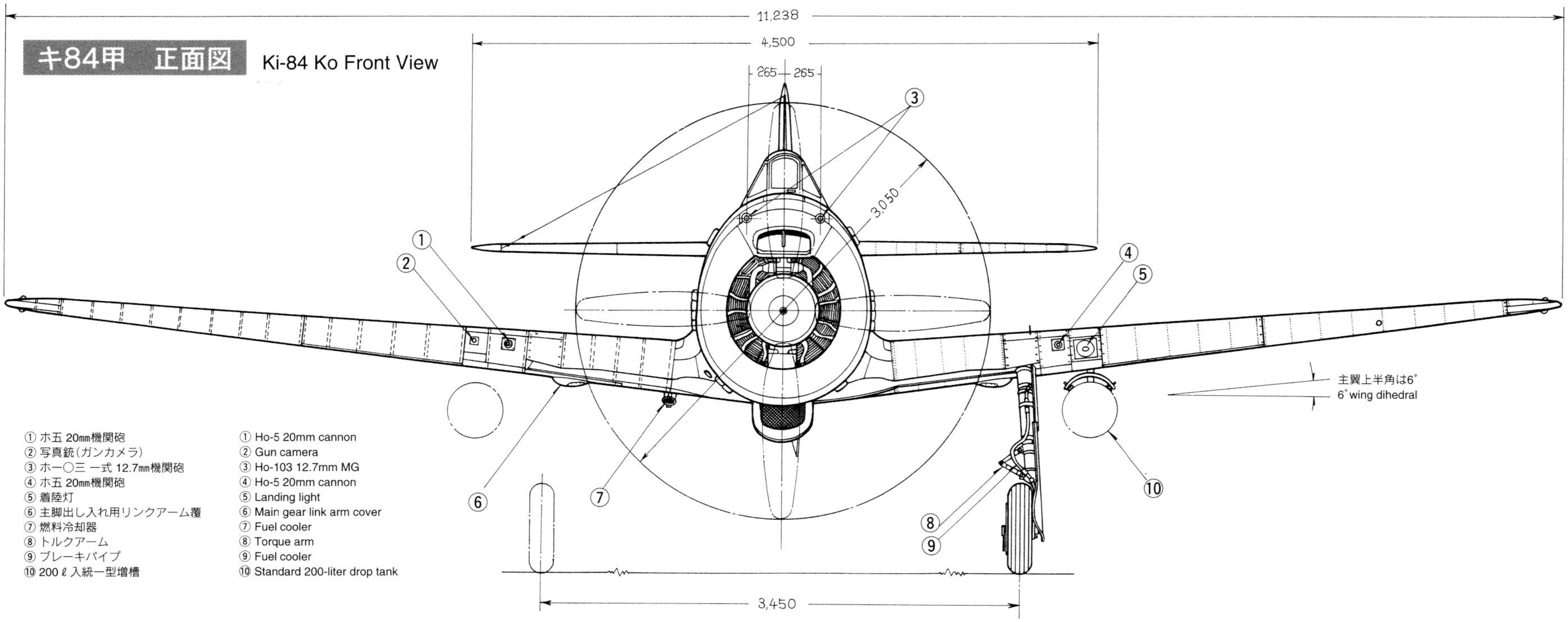

① ホ五 20㎜機関砲
② 写真銃(ガンカメラ)
③ ホー〇三 一式 12.7㎜機関砲
④ ホ五 20㎜機関砲
⑤ 着陸灯
⑥ 主脚出し入れ用リンクアーム覆
⑦ 燃料冷却器
⑧ トルクアーム
⑨ ブレーキパイプ
⑩ 200 ℓ 入統一型増槽

① Ho-5 20mm cannon
② Gun camera
③ Ho-103 12.7mm MG
④ Ho-5 20mm cannon
⑤ Landing light
⑥ Main gear link arm cover
⑦ Fuel cooler
⑧ Torque arm
⑨ Fuel cooler
⑩ Standard 200-liter drop tank

キ84甲 上面図　Ki-84 Ko Top View

① 翼端灯（赤）
② ピトー管
③ 着陸灯
④ 主脚引込装置点検パネル
⑤ ホ五 20㎜機関砲
⑥ 主翼内燃料タンク注入ハッチ
⑦ 点検パネル
⑧ ホ五点検ハッチ
⑨ 20㎜機関砲弾倉
⑩ 着脱式翼前縁燃料タンク
⑪ 燃料注入口ハッチ
⑫ 翼端灯（青）
⑬ 補助翼操作桿
⑭ 補助翼固定タブ
⑮ フラップ操作部点検ハッチ
⑯ フラップ作動ガイドレール
⑰ 昇降舵修正タブ操作部点検パネル
⑱ 前縁着脱作業用パネル
⑲ 昇降舵修正タブ

① Wingtip light (red)
② Pitot tube
③ Landing light
④ Landing gear mechanism access panel
⑤ Ho-5 20mm cannon
⑥ Mid-wing fuel tank fill hatch
⑦ Access panel
⑧ Ho-5 access hatch
⑨ 20mm magazine
⑩ Removable wing leading edge fuel tank
⑪ Fuel fill hatch
⑫ Wingtip light (blue)
⑬ Aileron control rod
⑭ Aileron fixed tab
⑮ Flap control access hatch
⑯ Flap guide rail
⑰ Elevator trim tab control access hatch
⑱ Leading edge removal access panel
⑲ Elevator trim tab

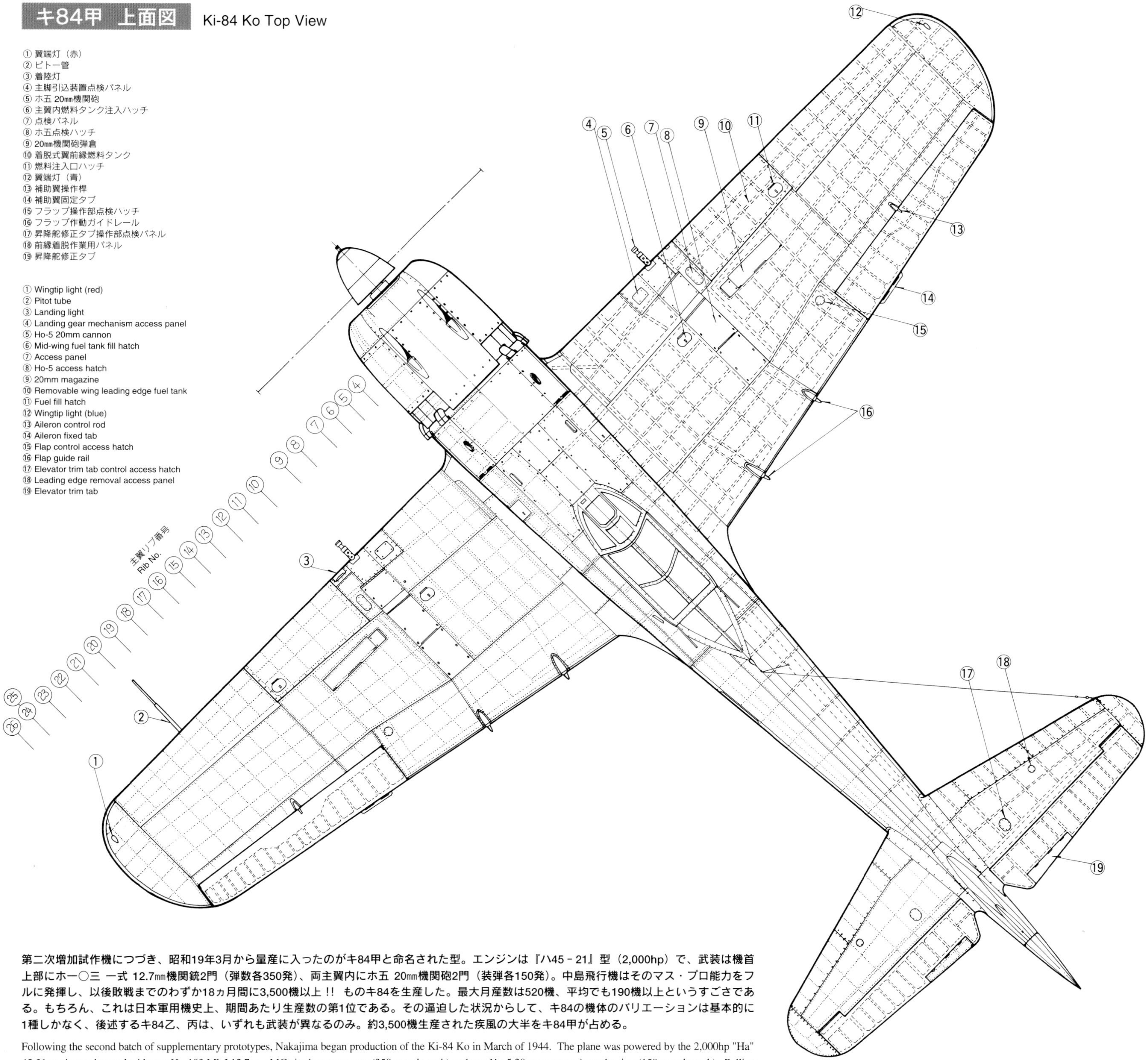

第二次増加試作機につづき、昭和19年3月から量産に入ったのがキ84甲と命名された型。エンジンは『ハ45－21』型（2,000hp）で、武装は機首上部にホ一〇三 一式 12.7㎜機関銃2門（弾数各350発）、両主翼内にホ五 20㎜機関砲2門（装弾各150発）。中島飛行機はそのマス・プロ能力をフルに発揮し、以後敗戦までのわずか18ヵ月間に3,500機以上！！ ものキ84を生産した。最大月産数は520機、平均でも190機以上というすごさである。もちろん、これは日本軍用機史上、期間あたり生産数の第1位である。その逼迫した状況からして、キ84の機体のバリエーションは基本的に1種しかなく、後述するキ84乙、丙は、いずれも武装が異なるのみ。約3,500機生産された疾風の大半をキ84甲が占める。

Following the second batch of supplementary prototypes, Nakajima began production of the Ki-84 Ko in March of 1944. The plane was powered by the 2,000hp "Ha" 45-21 engine and armed with two Ho-103 Mk.I 12.7mm MGs in the upper nose (350 rounds each) and one Ho-5 20mm cannon in each wing (150 rounds each). Pulling out all stops in their production capability, Nakajima turned out an amazing 3,500 Ki-84s in the 18 months between then and the end of the war. The represented an average production of 190 planes per month, and a peak of 520, truly remarkable figures for Japanese industry at the time. Of course, this record still stands as the most intensive period of aircraft production for any model in Japanese aviation history. Because of the needs of mass production, there is only one basic type of the Ki-84. The Otsu and Hei variations that followed the Ko differ in armament only. The Ko makes up well nearly 90% of the total Hayate production of 3,500.

キ84甲 下面図 Ki-84 Ko Underside

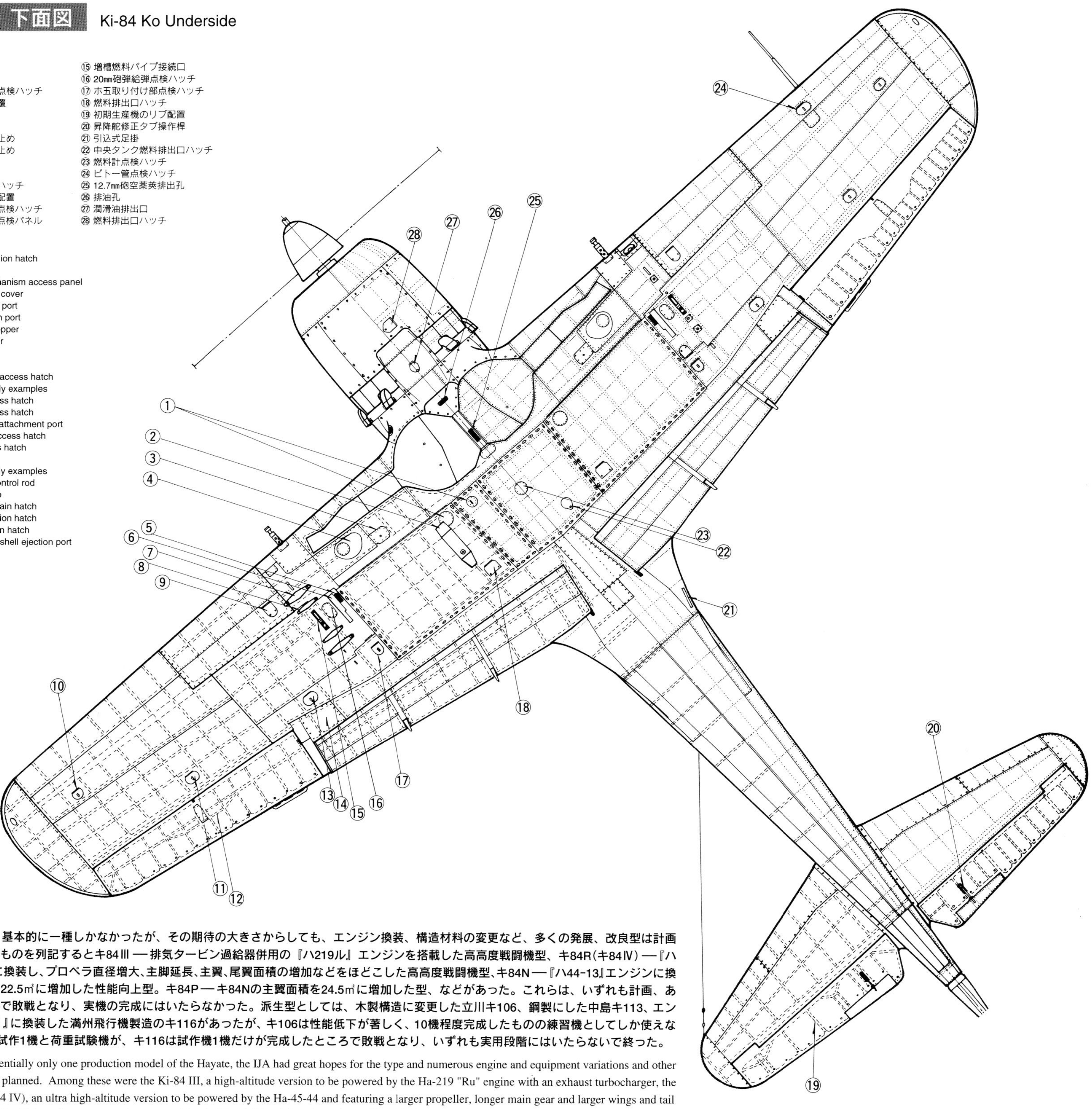

①燃料計点検ハッチ
②燃料冷却器
③主脚出し入れ装置点検ハッチ
④主脚リンクアーム覆
⑤保弾子排出口
⑥空薬莢排出口
⑦増槽懸吊時の振れ止め
⑧爆弾懸吊時の振れ止め
⑨燃料排出口
⑩点検ハッチ
⑪補助翼操作桿点検ハッチ
⑫初期生産機のリブ配置
⑬フラップ操作装置点検ハッチ
⑭フラップ操作装置点検パネル
⑮増槽燃料パイプ接続口
⑯20mm砲弾給弾点検ハッチ
⑰ホ五取り付け部点検ハッチ
⑱燃料排出口ハッチ
⑲初期生産機のリブ配置
⑳昇降舵修正タブ操作桿
㉑引込式足掛
㉒中央タンク燃料排出口ハッチ
㉓燃料計点検ハッチ
㉔ピトー管点検ハッチ
㉕12.7mm砲空薬莢排出孔
㉖排油孔
㉗潤滑油排出口
㉘燃料排出口ハッチ

① Fuel gauge inspection hatch
② Fuel cooler
③ Landing gear mechanism access panel
④ Main gear link arm cover
⑤ Empty clip ejection port
⑥ Spent shell ejection port
⑦ Drop tank sway stopper
⑧ Bomb sway stopper
⑨ Fuel drain
⑩ Access hatch
⑪ Aileron control rod access hatch
⑫ Rib location on early examples
⑬ Flap activator access hatch
⑭ Flap activator access hatch
⑮ Drop tank fuel line attachment port
⑯ 20mm shell feed access hatch
⑰ Ho-5 mount access hatch
⑱ Fuel drain hatch
⑲ Rib location on early examples
⑳ Elevator trim tab control rod
㉑ Retracting foot step
㉒ Central fuel tank drain hatch
㉓ Fuel gauge inspection hatch
㉔ Pitot tube inspection hatch
㉕ 12.7mm MG spent shell ejection port
㉖ Oil drain
㉗ Engine oil drain
㉘ Fuel drain hatch

キ84の量産型は、基本的に一種しかなかったが、その期待の大きさからしても、エンジン換装、構造材料の変更など、多くの発展、改良型は計画されていた。主なものを列記するとキ84Ⅲ―排気タービン過給器併用の『ハ219ル』エンジンを搭載した高高度戦闘機型、キ84R（キ84Ⅳ）―『ハ45-44』エンジンに換装し、プロペラ直径増大、主脚延長、主翼、尾翼面積の増加などをほどこした高高度戦闘機型、キ84N―『ハ44-13』エンジンに換装し、主翼面積を22.5㎡に増加した性能向上型。キ84P―キ84Nの主翼面積を24.5㎡に増加した型、などがあった。これらは、いずれも計画、あるいは設計の途中で敗戦となり、実機の完成にはいたらなかった。派生型としては、木製構造に変更した立川キ106、鋼製にした中島キ113、エンジンを『ハ-112Ⅱ』に換装した満州飛行機製造のキ116があったが、キ106は性能低下が著しく、10機程度完成したものの練習機としてしか使えなかった。キ113は試作1機と荷重試験機が、キ116は試作機1機だけが完成したところで敗戦となり、いずれも実用段階にはいたらないで終った。

While there was essentially only one production model of the Hayate, the IJA had great hopes for the type and numerous engine and equipment variations and other improvements were planned. Among these were the Ki-84 III, a high-altitude version to be powered by the Ha-219 "Ru" engine with an exhaust turbocharger, the Ki-84 R (a.k.a. Ki-84 IV), an ultra high-altitude version to be powered by the Ha-45-44 and featuring a larger propeller, longer main gear and larger wings and tail surfaces, the Ki-84 N, a high-performance version powered by the Ha-44-13 and featuring wings with 22.5㎡ more surface area, and the Ki-84 P, which boosted the wing area of the N by yet another 24.5㎡. These aircraft were all either in the planning or design stage at the time the war ended, and none ever were built. Variations that actually existed included the Tachikawa Ki-106, a wooden version of the plane, the steel-constructed Nakajima Ki-113, and the Manchu Aircraft Ki- 116, which was powered by the Ha-112-II. The Ki-106's performance deteriorated so severely that only 10 were built and these were used as trainers. Two Ki-113 prototypes and one Ki-116 prototype were completed at the time the war ended.

キ84乙 Ki-84 Otsu

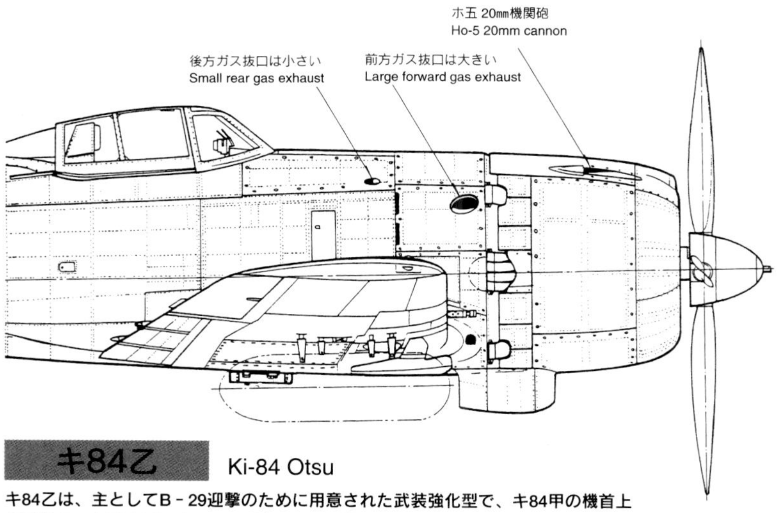

キ84丙 Ki-84 Hei

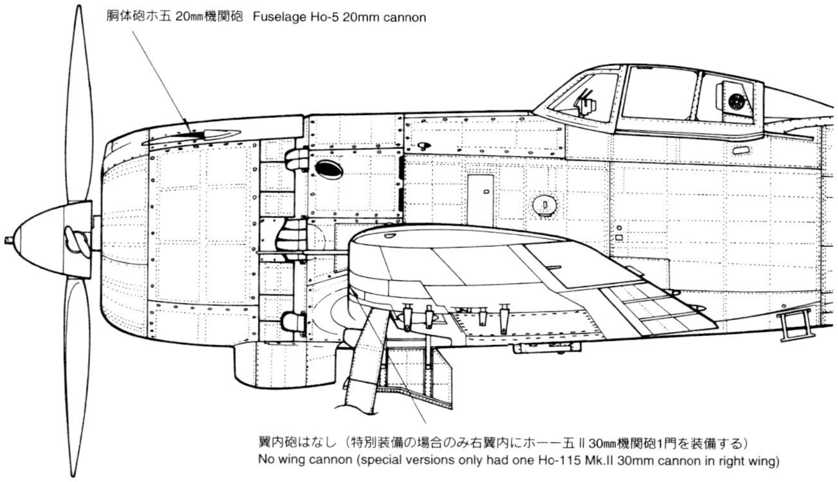

キ84乙 Ki-84 Otsu

キ84乙は、主としてB－29迎撃のために用意された武装強化型で、キ84甲の機首上部兵装を、主翼と同じホ五 20㎜機関砲に換装したもの。発射ガス抜き口の形状が異なるので、キ84甲との区別は容易。写真では、104戦隊への配備が確認できるが、数はそれほど多くなかったようだ。

The Ki-84 Otsu was a powerfully-armed variant of the plane designed primarily to knock down U.S. B-29s. The nose machine guns were changed to the same Ho-5 20mm cannon already in the wings of the Ko. The differently-sized gas exhaust ports in the nose makes distinguishing between the Ko and Otsu rather easy. Photographs testify that the type was used by the 104th FG, but it appears that its numbers were limited.

キ84丙 Ki-84 Hei

キ84丙も武装強化型で、キ84乙の主翼内にホー一五－II 30㎜機関砲各1門を装備可能とした。ただし、ホー一五－II は特別装備であり、通常は未整備が原則。ホー一五－II そのものが量産にいたらなかったため、試作程度の段階で終ったようだ。

The Ki-84 Hei was also a heavily-armed variant of the plane designed to be able to carry the Ho-115 30mm cannon in both wings. However, this armament was optional and rarely seen, and only in one wing when installed. The Ho-115 itself never reached mass production and the Hei never amounted to more than an experimental design as a result.

キ84丁 Ki-84 Tei

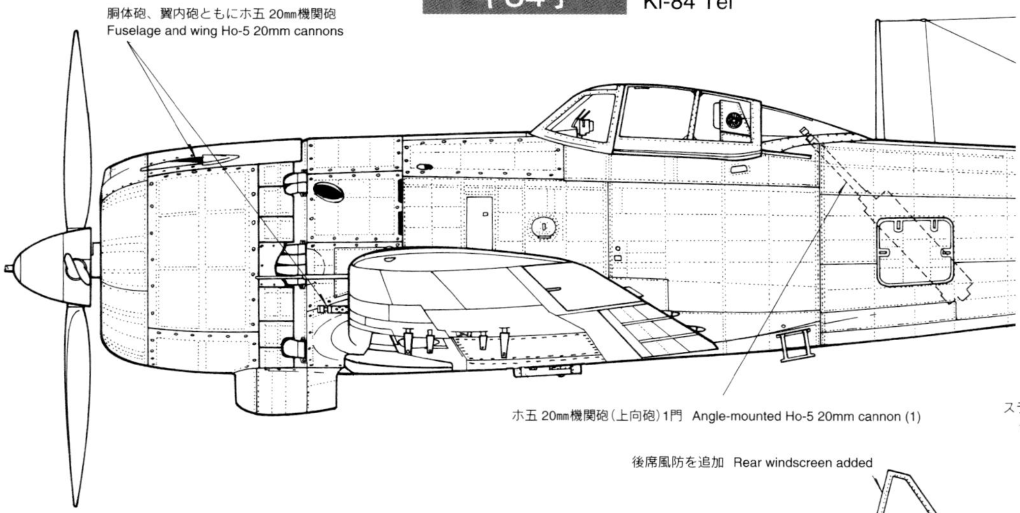

キ84丁 Ki-84 Tei

キ84丁はキ84乙の操縦室後方胴体内に、ホ五 20㎜機関砲1門を前上方に向けて追加装備した夜戦型。もっとも、単発戦闘機の夜間航法能力がきわめて貧弱だったことを考えると、実際に運用できたかどうかは疑問。試作機が2機つくられたのみ。

The Ki-84 Tei was a night-fighter version of the plane that featured a single Ho-5 20mm gun mounted at an upward angle behind the cockpit of the plane. Given that single-seat aircraft had extremely limited ability to navigate safely at night, it is doubtful whether this type could have been effective even if produced in quantity. Two prototypes were completed only.

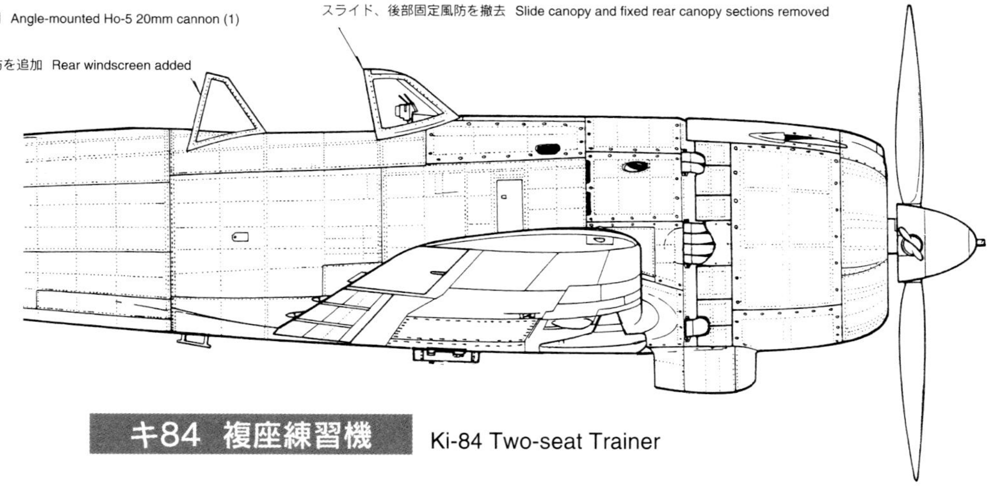

キ84 複座練習機 Ki-84 Two-seat Trainer

キ84 複座練習機 Ki-84 Two-seat Trainer

キ84のパイロット（陸軍では正式には空中勤務者と称した）を養成するため、生産ラインではなく、錬成飛行隊などの現地で複座に改造された練習機が少数存在した。前席は従来のスライド、後部固定風防を撤去、後席は簡単な固定風防を付けただけの、オープン・コクピットである。

In order to train pilots (the IJA called them "air servicemen") to fly the Ki-84, a few existing aircraft were converted to two-seat trainers in the field. These were open cockpit aircraft that had the sliding and fixed rear portions of the standard canopy removed, with a simple windscreen added behind the pilot's seat.

Aero Detail 24 Production Team

NAKAJIMA Ki-84 "FRANK" HAYATE

TEXT
野原 茂 Shigeru Nohara
田中理人 Masato Tanaka

PHOTOGRAPHS
野原 茂 Shigeru Nohara
(photos 9-23, 32-36, 38-39, 41-42, 45-51, 55, 59-64, 73, 78-83, 86-91, 102-122, 124, 126-130, 133-134, 139-145, 147-155, 157-159, page 5 center & bottom Page 7 top right)
潮書房 Ushio-shobo
(photos 1-8, 24-31,37, 40, 43-44, 52-54, 56-58, 65-72, 74-77, 84-85, 92-101, 123, 125, 131-132, 135-138, 146, 156, page 2, page 5 top, page 6 center & bottom, page 7 top left)
武田 明夫 Akio Takeda
(photo page 6 top)
知覧特攻平和記念館 Tokko Heiwa Kinen-kan,Chiran
(photo page 7 center right)

ILLUSTRATIONS
野原 茂 Shigeru Nohara

ENGLISH TEXT
スコット・ハーズ Scott T. Hards

COVER DESIGN
丹羽和夫 Kazuo Niwa

LAYOUT
佐藤 理 Osamu Sato

EDITOR
田中理人 Masato Tanaka
佐藤 理 Osamu Sato

SPECIAL THANKS TO :
潮書房 Ushio-shobo
武田 明夫 Akio Takeda
知覧特攻平和記念館 Tokko Heiwa Kinen-kan, Chiran
ジェイムズ・F・ランズデール James F. Lansdale

Japanese Pronunciation Guide

Though the long words can look intimidating, Japanese is one of the easiest languages in the world for speakers of western tongues to pronounce correctly. The whole language is structured around about just sixty different syllables that are either consanant/vowel, or single vowels, or a final 'n.' And since there are only five different vowel pronunciations, things are quite easy:

a as in "father"
e as in "bed"
i as in "pizza"
o as in "soap"
u as in "rude"

So a word like "hinomaru" breaks down as "hee-noh-mah-roo." Hayate is "hah-yah-teh." What could be easier?

Scott T. Hards (E-mail: scott@hlj.com)

In North America, the *Aero Detail* series, *Achtung Panzer* series, and other Model Graphix books are available from :
RZM IMPORTS
PO Box 995 Southbury, CT 06488 U. S. A.
Tel: 203-264-0774
Fax: 203-264-4967
Email: RZMIMPORT@aol.com

Quick Glossary
WWII Japanese Army Aviation Terms

Chutai (中隊): In IJA usage, a squadron of 9-12 aircraft.
Ha (ハ): Abbreviation for "hatsudoki" (発動機; engine). Used to preface engine model codes.
Hayabusa (隼): "Peregrine falcon." The nickname of the Type 1 Fighter (Ki-43), a.k.a. "Oscar."
Hayate (疾風): Generally translated as "gale" but the Japanese meaning also denotes a sudden, unwelcome, or ill-bringing wind. The nickname of the Type 4 Fighter (Ki-84), a.k.a."Frank."
Hien (飛燕): "Swallow." The nickname of the Type 3 Fighter (Ki-61), a.k.a."Tony." With a radial engine, it was known as the Type 5 Fighter (Ki-100).
Hinomaru (日の丸): Lit. "disk of the sun" but universally translated as "rising sun." Used to refer to the red national insignia circles (a.k.a. "meatballs") on Japanese planes. In modern Japanese, this expression usually denotes the country's flag.
Ho (ホ): Abbreviation for "hou (砲; gun)." Used to preface aircraft gun model codes.
Homare (誉): "Honor." Nickname for the Ha-45 series of engines.
Kamikaze / Shinpu (神風): "Divine Wind." Though originally referring to a sudden typhoon that saved Japan from Mongol invasion in the 13th century, it now usually refers to the suicide attacks used by Japanese air units in late WWII. In Japan, this tactic is referred to as "tokko," (特攻) or "special attacks."
Ki (キ): Translates as "craft," as in "aircraft." Used to preface IJA aircraft model codes.
Kinsei (金星): "Venus" (as in the planet; some sources mistranslate this as "gold star," a homonym). Nickname for the Ha-33 series of engines.
Sakae (栄): "Glory" or "prosperity." Nickname for the Ha-35 series of engines.
Sentai (戦隊): In IJA usage, and air group comprised of three chutais (中隊) and one HQ unit.
Shoki (鍾馗): "Devil queller." The nickname of the Type 2 Fighter/Interceptor (Ki-44), a.k.a. "Tojo."

〈編集後記〉

●エアロ・ディテールで日本機をとりあげるのは、No.7の零戦以来だから、ずいぶんと久しぶりなんだなぁー。四式戦に関してはボクの評価はかなり辛いのだが、べつに故意にそうしているわけではなく、実状をありのまま捉えているつもり。単に身びいきで必要以上に褒めそやすのは、かえって実機の価値を歪めてしまうことになる。
エンジンを別にすれば、たしかに四式戦の機体設計は、ソツなく合理的にまとまっていて、中島飛行機設計スタッフの実力のほどを示している。キ27、キ43、キ44と続いてきたいわゆる"中島流戦闘機設計持論"のひとつの到達点ということは理解できる。
戦時下の軍用機は、性能もさることながら、短期間にできるだけ多くつくれるといった、生産性の良さも欠かせない。わずか17カ月間に3,500機余も量産した中島飛行機のマス・プロ能力と、四式戦の生産性の高さは、その意味で大いなる誇りとしてよい。
それはともかくとして、1973年に千葉県の陸上自衛隊・木更津基地で、今回の主役である1446号機を初めて間近にみたときの感動は、今も鮮明に記憶に残っている。数日後に入間航空ショーで飛行したときの勇姿を。ナマで一目みたかったなぁー。もう二度とできないことだけに、かえすがえすも残念でならない。

野原 茂

●今回のエアロ・ディテールは、僕にとって初めてのことが多かった。日本機のため、いつものように僕が編集者のために元の資料を英訳するのではなく、逆に訳してもらわなければならない身になって、多くの面で担当者のお世話になったのだ。田中さん、ありがとう。
ところで、以前F4Fワイルドキャットに関しても書いたが、パイロットが良ければ、飛行機は最先端の優れものじゃなくても勝てる。疾風のことを勉強して、逆の意味でこの発想の正しさを痛感した。飛行機が優れものであっても、パイロットが50時間も飛んでいない新米じゃ、戦争には勝てません。戦果をあまり挙げられなかったのは、決してこのメカのせいではないことを確信している。

スコット・ハーズ

エアロ・ディテール24
中島 四式戦闘機「疾風」
発行日 1999年5月 初版第1刷

発行所 株式会社大日本絵画
〒101-0054 東京都千代田区神田錦町1丁目7番地
03-3294-7861 (代表)
編集 株式会社アートボックス
Edited by ART BOX Co., Ltd.
No.7, 1 Chome Kanda-Nishiki-cho Chiyoda-ku Tokyo
〒101-0054 Japan
印刷／製本 錦プロデューサーズ株式会社
Published May 1999 by Dai Nippon Kaiga Co., Ltd.
No.7, 1-chome Kanda-Nishiki-cho, Chiyoda-ku Tokyo
〒101-0054 Japan
© Copyright 1999 Dai Nippon Kaiga Co., Ltd.
Printed in Japan